SpringerBriefs in Molecular Science

Chemistry of Foods

Series editor

Salvatore Parisi, Industrial Consultant, Palermo, Italy

The series Springer Briefs in Molecular Science: Chemistry of Foods presents compact topical volumes in the area of food chemistry. The series has a clear focus on the chemistry and chemical aspects of foods, topics such as the physics or biology of foods are not part of its scope. The Briefs volumes in the series aim at presenting chemical background information or an introduction and clear-cut overview on the chemistry related to specific topics in this area. Typical topics thus include: - Compound classes in foods - their chemistry and properties with respect to the foods (e.g. sugars, proteins, fats, minerals, ...) - Contaminants and additives in foods - their chemistry and chemical transformations - Chemical analysis and monitoring of foods - Chemical transformations in foods, evolution and alterations of chemicals in foods, interactions between food and its packaging materials, chemical aspects of the food production processes - Chemistry and the food industry - from safety protocols to modern food production The treated subjects will particularly appeal to professionals and researchers concerned with food chemistry. Many volume topics address professionals and current problems in the food industry, but will also be interesting for readers generally concerned with the chemistry of foods. With the unique format and character of Springer Briefs (50 to 125 pages), the volumes are compact and easily digestible. Briefs allow authors to present their ideas and readers to absorb them with minimal time investment. Briefs will be published as part of Springer's eBook collection, with millions of users worldwide. In addition, Briefs will be available for individual print and electronic purchase. Briefs are characterized by fast, global electronic dissemination, standard publishing contracts, easy-to-use manuscript preparation and formatting guidelines, and expedited production schedules. Both solicited and unsolicited manuscripts focusing on food chemistry are considered for publication in this series.

More information about this series at http://www.springer.com/series/11853

Maria Micali · Marco Fiorino
Salvatore Parisi

The Chemistry of Thermal Food Processing Procedures

 Springer

Maria Micali
Industrial Consultant
Messina
Italy

Salvatore Parisi
Industrial Consultant
Palermo
Italy

Marco Fiorino
Technical Consultant
Siracusa
Italy

ISSN 2191-5407   ISSN 2191-5415   (electronic)
SpringerBriefs in Molecular Science
ISSN 2199-689X   ISSN 2199-7209   (electronic)
Chemistry of Foods
ISBN 978-3-319-42461-3   ISBN 978-3-319-42463-7   (eBook)
DOI 10.1007/978-3-319-42463-7

Library of Congress Control Number: 2016945103

Printed on acid-free paper

This Springer imprint is published by Springer Nature
The registered company is Springer International Publishing AG Switzerland

Contents

Chapter 1
Thermal Processing in the Food Industry. An Introduction

Maria Micali and Marco Fiorino

Abstract This chapter discusses briefly the connections between food degradation and correlated hygiene concerns on the one hand (reasons for food preservation), and available systems for the production of safe and legal food products on the other side (possible solutions). The reliable application of one or more of existing preservation and processing technologies—methods based on high or low temperatures; drying, salting, sugar addition and irradiation systems; the use of modified atmospheres; etc.—and the concomitant use of effective chemical additives may be helpful. However, the 'technological treatment/chemical preservation' two-binaries combination does not seem successful at present. Consequently, innovative technologies are also needed: these methods should be able to ensure the safety of food products and slow down degradation processes in foods and beverages, while preserving sensorial and nutritional properties.

Keywords Food degradation · Food preservation · Food safety · Freezing · Heating · Microbial spoilage · Refrigeration

Abbreviation

WHO   World Health Organization

1.1 Introduction to Thermal Processes for the Production of Safe Foods

Safety and hygiene of foods are always on the list of the most discussed topics, even in western Countries, despite recent advances in food technology and biotechnology (Wilcock et al. 2004). Recent analyses have shown that 30 % and more of the

M. Micali (✉)
Industrial Consultant, Messina, Italy
e-mail: marinamicali@virgilio.it

M. Fiorino
Technical Consultant, Siracusa, Italy
e-mail: marco.fiorino.ingegnere@gmail.com

© The Author(s) 2016
M. Micali et al., *The Chemistry of Thermal Food Processing Procedures*,
Chemistry of Foods, DOI 10.1007/978-3-319-42463-7_1

population in industrialised nations suffer from gastrointestinal disorders associated with the consumption of contaminated foods (WHO 2002); in addition, the estimation of shelf life for many food products can be erroneous or not easily demonstrable, with the augment of altered products and the consequent damage for food producers in terms of brand loyalty (Maloni and Brown 2006; Parisi 2002; Parisi et al. 2004). The reduction or elimination of undesired microorganisms in edible products for human consumption remains, therefore, a priority (the same reflection has to be considered when speaking of feeds for animal consumption).

The implementation of similar objectives could be simple enough. The reliable application of one or more of available preservation and processing technologies—treatments of thermal renovation; methods based on low temperatures; drying, salting, sugar addition and irradiation systems; the use of modified atmospheres—and the concomitant use of effective chemical additives may be helpful. However, the 'technological treatment/chemical preservation' two-binaries combination does not seem successful at present: the modern concept of perceived food quality, clearly identified with the Anglo-Saxon term 'green', cannot be easily correlated with non-artisanal food production strategies (Bonini and Oppenheim 2008; Boye and Arcand 2012; Young et al. 2010). In detail, the modern consumerist tendency in the Western world appears to move towards food products without a low (and demonstrable) content of chemical additives and a low impact on the environment; in other words, consumers' opinion is in favour of 'green' and 'low footprint' foods. In addition, the World Health Organization (WHO) has recently recommended the reduction of sodium chloride with the aim to reduce the risk and incidence of cardiovascular diseases. The recommendation may result in the increase in the use of chemical preservatives for many food products; certainly, the possible evolution does not reflect current consumers' preferences.

Consequently, innovative technologies are needed: these methods should be able to ensure the safety of food products, even in combination with existing ones, while preserving genuineness and 'naturalness'. In other terms, the reduction or elimination of chemical compounds in the preparation of foods is one of the major topics of current interest to the scientific world; on the other side, 'new' (evolved) foods and beverages should also assure the current performance in terms of safety, hygiene, shelf life expectations and legality. Current researches are focused on the study and development of new technologies and biotechnologies for conservation that are able to:

(a) Ensure food safety
(b) Influence positively sensory features of edible products
(c) Reduce the use of chemical additives
(d) Preserve nutritional features
(e) Maintain the original 'freshness' and 'naturalness' as perceived by the normal consumer (Parisi 2012).

Some studies have shown that the substitution of chemical additives with natural compounds may play an important role in future, on condition that natural additives (of vegetable, natural or microbial origin) can show an effective antimicrobial activity with the exclusion of adverse effects on consumers' health (Cleveland et al.

2001; de Silva 1995; Gachkar et al. 2007; Lanciotti et al. 2004; Smid and Gorris 2007). These actions have found a strong stimulus even in the fields of funded research, in accordance with the last two Framework Programmes in the European Union (Aguilar et al. 1998). In this ambit, several of these natural compounds—nutraceutical substances—have been found to preserve qualitative food features and exert positive effects on consumers' health at the same time (Sumi 2008; Summet et al. 2010). These approaches may promote the reduction (or elimination, when possible) of chemical additives in food preparations, and improve nutritional features in synergy with traditional systems for the reliable preservation of foods (Bhat et al. 2012; Zhang et al. 2011).

1.2   The Safety of Foods

Basically, a healthy food does not contain pathogens (protozoa, bacteria, viruses, prions) or microbiological toxins (D'Mello 2003; Fletcher et al. 2012; Lindenbaum 2001; Tauxe 2001); it is not infested with micro-parasites such as *Taenia solium*, *T. saginata*, *Trichinella spiralis*, *Echinococcus granulosus*, *Anisakis* spp, etc. (Macpherson 2005; Orlandi et al. 2002) and does not contain poisons that pollute the natural habitat. Moreover, no contaminations have to be observed after contact with non-edible substances such as utensils (dishes and containers) or 'virgin' and 'recycled' food packaging materials (Brunazzi et al. 2014; Bryan 1992; Parisi 2012, 2013). In addition, contamination substances include pesticide residues, undesired or unexpected additives, antibiotics, hormones and similar substances, radioactivity and spoilage microorganisms (Donoghue 2003; Epstein 1990; Khaniki 2007). Finally, foods and beverages may be judged unsafe if a selected and allowed additive (carbon monoxide, glutamate, nicotinic acid, niacin, etc.) may be found in excessive concentrations, probably because of wrong dosages or the non-uniform distribution in the intermediate mass during the preparation (Rozan et al. 2000).

1.3   Food Preservation

Food preservation is crucial when speaking of food hygiene and health: each single preservation treatment is mainly designed with the aim of delaying the inevitable alteration of quickly perishable foods between production and consumption. In addition, the distribution of food commodities is favoured in different geographical areas and temporal periods of the year; for these reasons, economic implications of food preservation are evident (Maroulis and Saravacos 2007; Zanoni and Zavanella 2012).

By a statistical point of view, food alterations are mainly due to the development of different spoilage microorganisms, with the additional presence of potential pathogen agents, and enzyme-mediated reactions, provided that a certain amount of bioavailable water is present (Chap. 3). Consequently, the following two actions have to be carried out if the final aim is the increase of food and beverage stability:

(1) The destruction of pathogen agents (matter of public hygiene and safety) and
     the reduction of spoilage microorganisms (visible effects on foods)
(2) The inactivation of microbial enzymes, if already synthesised.

These results can be achieved by means of systems such as physical techniques—
heating, refrigeration, freezing, drying, irradiation—or biological strategies
(fermentation). Anyway, some general conditions have to be always satisfied:

- The food items must be of good quality
- The chosen treatment has to be carried on as early as possible to avoid the onset
  of microbial or enzymatic alterations
- The manipulation of raw materials and intermediate food masses have to be
  minimised because each step or sup-step without a clear preservation effect may
  potentially increase microbial contamination
- Preliminary washing operations are extremely useful in certain situations (e.g.
  fresh vegetable raw materials) if high-level superficial contamination is
  demonstrated and/or forecasted.

Food preservation means an evolving number of techniques that are used to slow
down degradation processes in foods and beverages, and preserve sensorial and
nutritional properties. Generally, these methods imply the creation of unfavourable
conditions to the development, activity and life of various pathogens, spoilage
microorganisms and other life forms. Used treatments tend to preserve the integrity
and the legality of foods with different mechanisms.

Traditional preservation treatments are substantially smoking, keeping in vinegar, preservation in oil or alcohol, drying, salting, fermentation and preservation in
sugar (Rahman 2007). High-temperature methods include pasteurisation, sterilisation and similar techniques, while approaches based on low-temperature storages
(Gould 1995; Leistner 1992) are refrigeration, and freezing (different sub-
categories). Irradiation is a powerful technique when used in combination with
other preservation approaches (Lacroix and Ouattara 2000). Finally, preservation
systems based on water removal are dehydration, irradiation, lyophilisation and
concentration, while the use of chemical preservatives has to be considered as a
group apart. Chapter 2 describes the above-mentioned approaches in detail.
Different systems may be also used at present, including 'pulsed electric fields
technology', ultrasounds and high-pressure treatments (Chemat et al. 2011; Heinz
and Buckow 2010; Lelieveld et al. 2007; Toepfl et al. 2007; Vega-Mercado et al.
1997), but these technologies are not described in this book.

References

Aguilar A, Ingemansson T, Magnien E (1998) Extremophile microorganisms as cell factories:
    support from the European Union. Extremophiles 2(3):367–373. doi:10.1007/s007920050080
Bhat R, Karim Alias A, Paliyath G (eds) (2012) Progress in food preservation. Wiley, Chichester.
    doi:10.1002/9781119962045

Bonini S, Oppenheim J (2008) Cultivating the green consumer. Stanf Soc Innov Rev 6(4):56–61

Boye JI, Arcand Y (2012) Green technologies in food production and processing. Food Engineering Series, Springer Science & Business Media. doi:10.1007/978-1-4614-1587-9

Brunazzi G, Parisi S, Pereno A (2014) Packaging and quality. In: Brunazzi G, Parisi S, Pereno A (eds) The importance of packaging design for the chemistry of food products. Springer International Publishing, Cham. doi:10.1007/978-3-319-08452-7_5

Bryan FL (1992) Hazard analysis critical control point evaluations: a guide to identifying hazards and assessing risks associated with food preparation and storage. World Health Organization, Geneva. http://apps.who.int/iris/handle/10665/37314. Accessed 26 Apr 2016

Chemat F, Zill-e-Huma, Khan MK (2011) Applications of ultrasound in food technology: processing, preservation and extraction. Ultrason Sonochem 18(4):813–835. doi:10.1016/j.ultsonch.2010.11.023

Cleveland J, Montville TJ, Nes IF, Chikindas ML (2001) Bacteriocins: safe, natural antimicrobials for food preservation. Int J Food Microbiol 71(1):1–20. doi:10.1016/s0168-1605(01)00560-8

de Silva KT (1995) A manual on the essential oil industry. United Nations Industrial Development Organization, Wien. https://www.unido.org/fileadmin/user_media/Publications/Pub_free/A_manual_on_the_essential_oil_industry.pdf. Accessed 26 Apr 2016

D'Mello JPF (ed) (2003) Food safety: contaminants and toxins. CABI Publishing, Oxon and Cambridge. doi:10.1079/9780851996073.0000

Donoghue DJ (2003) Antibiotic residues in poultry tissues and eggs: human health concerns? Poult Sci 82(4):618–621. doi:10.1093/ps/82.4.618

Fletcher SM, Stark D, Harkness J, Ellis J (2012) Enteric protozoa in the developed world: a public health perspective. Clin Microbiol Rev 25(3):420–449. doi:10.1128/CMR.05038-11

Epstein SS (1990) Potential public health hazards of biosynthetic milk hormones. Int J Health Services 20(1):73–84

Gachkar L, Yadegari D, Rezaei MB, Taghizadeh M, Astaneh SA, Rasooli I (2007) Chemical and biological characteristics of Cuminum cyminum and Rosmarinus officinalis essential oils. Food Chem 102(3):898–904. doi:10.1016/j.foodchem.2006.06.035

Gould GW (ed) (1995) New methods of food preservation. Springer Science & Business Media, B.V., Dordrecht

Heinz V, Buckow R (2010) Food preservation by high pressure. J Verbraucherschutz Lebensmittelsicherh 5(1):73–81. doi:10.1007/s00003-009-0311-x

Khaniki GR (2007) Chemical contaminants in milk and public health concerns: a review. Int J Dairy Sci 2(2):104–115. doi:10.3923/ijds.2007.104.115

Lacroix M, Ouattara B (2000) Combined industrial processes with irradiation to assure innocuity and preservation of food products—a review. Food Res Int 33(9):719–724. doi:10.1016/S0963-9969(00)00085-5

Lanciotti R, Gianotti A, Patrignani F, Belletti N, Guerzoni M, Gardini F (2004) Use of natural aroma compounds to improve shelf-life and safety of minimally processed fruits. Trends Food Sci Technol 15(3–4):201–208. doi:10.1016/j.tifs.2003.10.004

Leistner L (1992) Food preservation by combined methods. Food Res Int 25(2):151–158. doi:10.1016/0963-9969(92)90158-2

Lelieveld HLM, Notermans S, De Haan SWH (eds) (2007) Food preservation by pulsed electric fields: from research to application. Woodhead Publishing Ltd, Cambridge; CRC Press LLC, Boca Raton

Lindenbaum S (2001) Kuru, prions, and human affairs: thinking about epidemics. Annu Rev Anthropol 30(1):363–385. doi:10.1146/annurev.anthro.30.1.363

Macpherson CNL (2005) Human behaviour and the epidemiology of parasitic zoonoses. Int J Parasitol 35(11):1319–1331. doi:10.1016/j.ijpara.2005.06.004

Maloni MJ, Brown ME (2006) Corporate social responsibility in the supply chain: an application in the food industry. J Bus Eth 68(1):35–52. doi:10.1007/s10551-006-9038-0

Maroulis ZB, Saravacos GD (2007) Food plant economics. CRC Press LLC, Boca Raton. doi:10.1201/9781420005790

Orlandi PA, Chu DMT, Bier JW, Jackson GJ (2002) Parasites and the food supply. Food Technol 56(4):72–79

Parisi S (2002) I fondamenti del calcolo della data di scadenza degli alimenti: principi ed applicazioni. Ind Aliment 41(417):905–919

Parisi S (2012) Food packaging and food alterations. Smithers Rapra Technology Ltd, Shawbury

Parisi S (2013) Food Industry and packaging materials: performance-oriented guidelines for users. Smithers Rapra Technology Ltd, Shawbury

Parisi S, Delia S, Laganà P (2004) Il calcolo della data di scadenza degli alimenti: la funzione Shelf Life e la propagazione degli errori sperimentali. Ind Aliment 43(438):735–749

Rahman MS (ed) (2007) Handbook of food preservation, 2nd edn. CRC Press LLC, Boca Raton

Rozan P, Kuo YH, Lambein F (2000) Free amino acids present in commercially available seedlings sold for human consumption. A potential hazard for consumers. J Agric Food Chem 48(3):716–723. doi:10.1021/jf990729v

Smid EJ, Gorris LGM (2007) Natural antimicrobials for food preservation. In: Rahman MS (ed) Handbook of food preservation, 2nd edn. CRC Press LLC, Boca Raton

Sumi Y (2008) Research and technology trends of nutraceuticals. Sci Technol Trends Q Rev 28:10–21. http://data.nistep.go.jp/dspace/bitstream/11035/2782/1/NISTEP-STT028E-10.pdf. Accessed 27 Apr 2016

Summet G, Devesh C, Kritika M, Preeti S, Anroop N (2010) An overview of nutraceuticals: current scenario. J Basic Clin Pharm 1(2):55–62

Tauxe RV (2001) Food safety and irradiation: protecting the public from foodborne infections. Emerg Infect Dis 7(7):516–521. doi:10.3201/eid0707.017706

Toepfl S, Heinz V, Knorr D (2007) High intensity pulsed electric fields applied for food preservation. Chem Eng Process Process Intensif 46(6):537–546. doi:10.1016/j.cep.2006.07.011

Vega-Mercado H, Martin-Belloso O, Qin BL, Chang FJ, Góngora-Nieto MM, Barbosa-Cánovas GV, Swanson BG (1997) Non-thermal food preservation: pulsed electric fields. Trends Food Sci Technol 8(5):151–157. doi:10.1016/S0924-2244(97)01016-9

WHO (2002) Food safety and foodborne illness. World Health Organization Fact sheet N° 237, Revised March 2007. World Health Organization, Geneva

Wilcock A, Pun M, Khanona J, Aung M (2004) Consumer attitudes, knowledge and behaviour: a review of food safety issues. Trends Food Sci Technol 15(2):56–66. doi:10.1016/j.tifs.2003.08.004

Young W, Hwang K, McDonald S, Oates CJ (2010) Sustainable consumption: green consumer behaviour when purchasing products. Sustain Dev 18(1):20–31. doi:10.1002/sd.394

Zanoni S, Zavanella L (2012) Chilled or frozen? Decision strategies for sustainable food supply chains. Int J Prod Econ 140(2):731–736. doi:10.1016/j.ijpe.2011.04.028

Zhang HQ, Barbosa-Cánovas GV, Balasubramaniam VB, Dunne CP, Farkas DF, Yuan JT (eds) (2011) Nonthermal processing technologies for food, vol 45. Wiley, Chichester

Chapter 2
Thermal Processing in Food Industries and Chemical Transformation

Maria Micali and Marco Fiorino

Abstract This chapter reviews thermal processes in the food industry—pasteurisation, sterilisation, UHT processes and others. It evaluates the effects on a chemical level and possible failures from a safety viewpoint and discusses in how far the effects can be predicted. In addition, historical preservation techniques—smoking, addition of natural additives, irradiation, etc.—are compared with current industrial systems, like fermentation, irradiation and addition of food grade chemicals. This chapter critically discusses storage protocols—cooling, freezing, etc.—and packing systems (modified atmosphere technology, active and intelligent packaging). Can undesired chemical effects on the food products be reliably predicted? This chapter elucidates on this important question. On that basis, new challenges, which currently arise in the food sector, can be approached.

Keywords Freezing · Modified atmosphere · Pasteurisation · Refrigeration · Semi-preserved food · Sterilisation

Abbreviations

CAS   Controlled atmosphere storage
HTST   High-temperature short-time
HHST   Higher heat shorter time
HPP   High-pressure processing
IQF   Individual quick freezing
MTA   Minimum temperature accretion
MAP   Modified atmosphere packaging

M. Micali (✉)
Industrial Consultant, Messina, Italy
e-mail: marinamicali@virgilio.it

M. Fiorino
Technical Consultant, Siracusa, Italy
e-mail: marco.fiorino.ingegnere@gmail.com

© The Author(s) 2016
M. Micali et al., *The Chemistry of Thermal Food Processing Procedures*,
Chemistry of Foods, DOI 10.1007/978-3-319-42463-7_2

RH   Relative humidity
TTI   Time temperature indicator
UHT   Ultra-High Temperature
a_w Water activity

2.1 Introduction to Food Preservation Methods. High-Temperature Systems

Food preservation is crucial when speaking of food hygiene and health: each single preservation treatment is mainly designed with the aim of delaying the inevitable alteration of quickly perishable foods between production and consumption. Generally, food alterations are considered in relation with microbial spoilage, with the additional presence of potential pathogen agents. In addition, enzymatic reactions have an important 'weight' when speaking of food alterations, on condition that a certain amount of bioavailable water is present. As a result, food preservation should be designed and performed with two basic aims at least:

- The destruction of pathogen agents and the reduction of spoilage microorganisms
- The inactivation of microbial enzymes, if already synthesised.

The chosen preservation strategy (or strategies) should satisfy the following conditions:

- Treated food items have to be optimal, when speaking of microbial counts (low numbers) and freshness
- The respect of times is crucial. Treatments have to be carried on as early as possible to avoid the onset of microbial or enzymatic alterations
- Manipulation and general processing steps should be minimised because each step or sup-step without a clear preservation effect may be cause of microbial contamination, recontamination or spoilage
- Preliminary washing operations are performed when needed, if high-level superficial contamination is demonstrated and/or forecasted.

Food preservation methods imply the creation of unfavourable conditions to the development, activity and life of various pathogens, spoilage microorganisms and other life forms. On the other side, treated foods have to be and remain constantly safe and legally compliant until the end of labelled durability periods.

This chapter is dedicated to the description of many preservation systems: Sect. 2.1 discusses high-temperature methods, including pasteurisation, sterilisation and similar techniques. Subsequently, approaches based on low-temperature storages (refrigeration and freezing) are discussed. Finally, this chapter considers traditional preservation treatments and systems based on water removal. Different techniques may be also used at present, including pulsed electric fields technology, ultrasounds and high-pressure treatments, but these technologies are not described in this book.

2.1.1 Pasteurisation: General Features

Basically, two different food preservation categories are performed with heat application: pasteurisation and sterilisation (Rahman 2007). The first method of preservation is defined which is able to destroy pathogenic life forms, and the most part of those vegetative microorganisms in foods (Alais 1984); moreover, enzymes are reported to be inactivated with pasteurisation treatments. Historically, the process is ascribed to Louis Pasteur in relation to his experiences in 1863 with 60 °C—treated wines and ameliorated shelf life periods (Alais 1984; Fry 2001; Rogers 2013).

In detail, pasteurisation has to be described as a microbiostatic and partially microbicidal treatment; for these reasons, it has to be sinergically accompanied with other preservation methods such as refrigeration, addition of chemical preservatives, or vacuum packaging (Andersen et al. 1990; Degraeve et al. 2003; Mangalassary et al. 2008; Murphy et al. 2006). Anyway, heat treatments should be followed by the rapid cooling of treated products with the aim of avoiding storage temperatures that could favour enzyme-catalysed or simple chemical reactions. Generally, processing times depend on the nature and typology of foods: at present, low-temperature pasteurisation may be used when speaking of beers and wines, while milk for cheesemaking purposes must be heated at 75 °C for 15 s. On the other hand, high-temperature short-time (HTST) and high-temperature pasteurisation techniques are used for milk and fresh milk because of higher thermal values and lower processing times (Table 2.1).

Table 2.1 A brief overview of pasteurisation and sterilisation processes

Pasteurisation and sterilisation processes	Minimum temperature, °C	Maximum temperature, °C	Processing times
Low-temperature pasteurisation	60	65	30 min
High-temperature pasteurisation	75	85	≤3 min
High-temperature short-time (HTST) pasteurisation	75	85	≤20 s
Classical sterilisation (appertisation)	100	120	>20 min
Indirect UHT system	140	150	<10 s
Direct UHT system	140	150	<10 s

Low-temperature pasteurisation may be used for beers and wines, while milk for cheesemaking purposes is expected to be heated at 75 °C at least for 15 s. HTST and high-temperature pasteurisation techniques are used for milk and fresh milk. Appertisation can be used when speaking of canned vegetables and animal products. Indirect and direct UHT systems are recommended for aseptic packaged foods (UHT milk, soups, baby foods, etc.); direct UHT is preferred for low-viscosity products because of the use of injection or infusion techniques

Four different pasteurisation treatments are currently used in the food and beverage industry (Alais 1984; Murphy et al. 2003; Rahman 2007):

(a) In-package pasteurisation. Temperature gradients can be used in certain situations. Fluids can be treated directly into bottles or cans. The process may be explained in the following way: the fluid is initially placed in the 'cold zone' of the container at about 2.2 °C; subsequently, it continues its diffusion towards hotter areas of the containers with a final 60 °C value. After a certain period at 60 °C, the fluid is progressively cooled until the final temperature is between 21.1 and 26.7 °C (Engelman and Sani 1983). In-package systems can be used with reliable effects on solid products such as ready-to-eat meat products with the aim of destroying pathogens and reducing post-process contamination caused by handling (Mangalassary et al. 2004)

(b) Pasteurisation before the final packaging. High-temperature gradients are recommended when speaking of thermally sensible foods. For this reason, preheating stages are a useful option. Products such as processed juices and spreadable butter-like compositions may be treated with this system (Ahmed and Luksas 1988; Samani et al. 2016)

(c) 'Batch' pasteurisation, also named low-temperature short-time pasteurisation. This system is specifically designed for fluids (milk, orange and apple juices, liquid whole eggs, etc.): the intermediate food is pasteurised at 62.8 °C for 30 min (Brant et al. 1968; Cinquanta et al. 2010; Cunningham 1986; Potter and Hotchkiss 1995; Wheeler et al. 1987). Interestingly, this treatment has been also used for waste raw materials into farms with good results when speaking of peculiar microbial contamination (Stabel 2001)

(d) Continuous or HTST pasteurisation (Sect. 2.1.1.1). Fluids are treated at 71.7 °C at least for 15 s or more (heat exchange is required). This method is discussed in Sect. 2.1.1.1.

2.1.1.1 Pasteurisation: The HTST Method

The HTST method, also known as 'flash' pasteurisation, is performed (Rigo 2010) on liquid foods passing through a two-heated walls system (heat exchanger plates) with an empty space of 1 mm thickness only (temperatures between 75 and 85 °C, times: 15–20 s). Heat exchange occurs immediately in a uniform manner, while germs tend to adhere to surfaces (wall effect). Another similar system, ascribed to Stassano, implies the following time and temperature conditions: 14 s, 75 °C (Rigo 2010). Anyway, rapid cooling is required after heating (International Dairy Foods Association 2006). An interesting variation is the Higher Heat Shorter Time (HHST) system; it requires different machines and slightly different conditions (temperature >89 °C, holding time: one second or less) with consequent higher durabilities and enhanced enzymatic inhibition (Chandan 2015; International Dairy Foods Association 2006). Should HHST system be used, treated products would be sometimes defined 'extended shelf life' foods. The extreme solution is represented by ultra-pasteurisation systems with temperature equal to 138 °C (minimum value) for at least 2 s (Chandan 2015). Shelf life values are enhanced: however, refrigeration is always required.

2.1.2 Sterilisation: General Features

Sterilisation is a more drastic treatment of food and non-food products, if compared with pasteurisation. The main difference is represented by the desired objective: the destruction of all microbial forms, including spores. Sterilisation may be defined as a commercial heat treatment, which can be able to destroy all living microorganisms in food during storage and distribution. On the other side, treated foods cannot be defined completely 'aseptic'; additionally, organoleptic and nutritional characteristics are expected to suffer some alteration. It should be also noted that treatments can differ according to the pH of preserved foods (Sect. 2.3): in detail, intermediate foods with pH values < 4.5 should require temperatures around 100 °C, while products with pH > 4.5 would need at least 120 °C for 20 min (minimum time). In general, there are three main sterilisation treatments with different conditions, according to Table 2.1. These methods—classic sterilisation or appertisation, indirect and direct ultra-high temperature (UHT) systems—are discussed in Sects. 2.1.2.1 and 2.1.2.2.

2.1.2.1 Appertisation

Historically, the 'appertisation' method—ascribed to Nicholas Francois Appert (1804)—has been considered a low sterilisation procedure because of the use of open baths with temperatures ≤ 100 °C (Borde 2006): foods were packaged in sealed glass containers. At present, appertisation is performed with autoclaves, pressurised chambers with boilers for the production of saturated steam (temperature >100 °C) and devices allowing the modification of temperature and pressure values. Moreover, the evolution of used containers—metal packages above all—and the enhanced thermal conductivity have to be taken into account. For these reasons, the classical sterilisation can be very useful when speaking of packaged solid foods; liquid foods are usually pasteurised (Larousse 1991).

2.1.2.2 UHT Pasteurisation

The UHT method differs from the classical sterilisation because it reaches rapidly higher temperatures with very short duration (direct UHT system); as a result, organoleptic and nutritional properties of treated foods should suffer a few variations only. The UHT method also applies to bulk products (heat penetration is extremely rapid) and is associated with aseptic packaging techniques.

Two different UHT systems can be used (Andreini et al. 1990; Clare et al. 2005; Morales et al. 2000; Rerkrai et al. 1987):

(a) Indirect method. In detail, thermal energy is supplied by means of an indirect steam injection system; fluids are forced to pass through close heat exchanger plates. It should be noted that thermal energy is diffused from the outside zone of UHT chambers to the inner 'core' (Corzo et al. 1994). As a consequence, external areas of the ideally considered fluid food may be superheated with possible variations (protein profiles, emersion of 'cooked' flavours because of the production of volatile sulphur components, etc.)

(b) Direct system. The preheated food and the injected heating medium (steam) are in direct contact for a few seconds; subsequently, under-vacuum cooling is required, similarly to pasteurisation treatments, with the aim of eliminating extra water (a certain volumetric augment is observed in the process). The reduced time of contact explains apparently the absence of serious damages, from the sensorial viewpoint (Datta et al. 2002).

A brief comparison of indirect and direct UHT methods seems to highlight the importance and advantages of the direct injection: sensorial variations seem reduced; on the other hand, these techniques may cause localised burns. An interesting 'infusion' steam version—the fluid food is infused into steam—is reported to minimise this defect (Datta et al. 2002).

Nutritional variations may be observed; on the other side, it may be affirmed that digestibility is improved. For these reasons, different non-heat-based systems (freezing methods) are apparently preferred at present.

2.2   Cold Preservation

This method of food preservation is based on the ability of cold to interfere with the vital functions of microorganisms: essentially, cold systems can assure a reliable microbiostatic action.

The cold slows down until it stops degradative (chemical and enzymatic) reactions inside the food. This effect is mainly due to the inhibition of microbial enzymes: they are not denatured and therefore inactivated, but their action is inhibited (rendered useless) in relation to the low temperature of the food. Moreover, the inhibitory action of low temperatures occurs with the subtraction of free water from the system because of ice formation with the consequent increase in the concentration of extracellular solutions and the dehydration of microbial cells.

The minimum temperature accretion (MTA) of microorganisms is needed in order to effectively maintain cold conditions. MTA of pathogenic microorganisms is always >0 °C, while psychrophilic life forms remain viable at lower temperatures. The microbial growth can be considered negligible at −18 °C and lower thermal values. Moreover, cold has a partial microbicidal action: a number of microbes suffer the irreversible alteration of cytoplasmic proteins and mechanic damages (formation of ice crystals) during the early stages of freezing treatments; however, surviving life forms may be ready to resume vital functions with spores and toxins, if thermal values are increased. For these reasons, foods should have excellent or good hygiene conditions before treatments.

2.2.1   Cold Storage: Different Preservation Techniques

2.2.1.1   Refrigeration

There are two different methods of food preservation by cold storage (Fig. 2.1): refrigeration and freezing systems. The first method, refrigeration preservation, concerns the storage of food commodities at temperatures low enough to slow down degradative (chemical and enzymatic) reactions inside stored products; in addition, water must remain liquid.

The choice of storage temperatures depends on the type of food and the required retention time: anyway, storage times are not excessive. Generally, refrigeration temperatures are between −1 and 8 °C. However, refrigerated storage shows good performances on condition that other factors are controlled; ventilation has to be

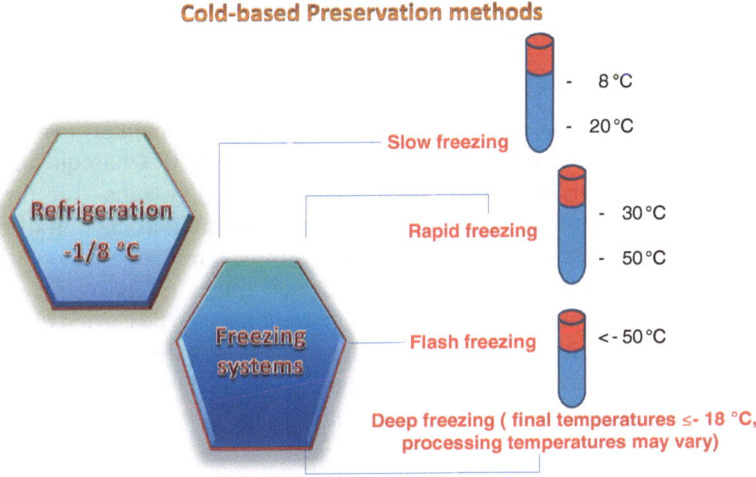

Fig. 2.1 A brief overview of cold preservation techniques. The first method, refrigeration, concerns the storage of food commodities at temperatures low enough to slow down degradative reactions inside stored products; in addition, water must remain liquid. Usually, temperatures remain between −1 and 8 °C. On the other side, freezing systems involve the exposure of foods and beverages to low or very low temperatures with the consequent water crystallisation and the solidification of the whole product. Four main systems—slow, rapid, flash and deep freezing—are used at present with different temperatures and processing times

assured into storage chills. Moreover, acceptable relative humidity (RH) values are between 80 and 95 %: reduced RH values may cause excessive water evaporation from food surfaces, while excessive RH may favour the development of moulds (visible mycelia).

Anyway, refrigerated storage can be used in different situations and for the following food categories at least:

(a) Refrigerated products during transport and storage only (sold at room temperature): fruits, vegetables, eggs
(b) Edible products distributed, stored and sold between −1 and 8 °C
(c) Prepared and precooked foods: salads, sandwiches, fresh pastries, precooked and delicatessen products, meats and cold cuts.

Consequently, the diffusion of refrigeration systems has been progressively introduced everywhere in the last decades with excellent results (Bruhn and Schutz 1999), despite several problems may be observed when speaking of food safety practices in consumers' homes (Karabudak et al. 2008; Redmond and Griffith 2003).

Historically, the use of natural and artificially produced snow and ice was the first homemade application of food cold storage. With reference to industrial applications, refrigeration was mainly based on the liquefaction by compression of several allowed gases such as ammonia, sulphur dioxide, carbon dioxide and chloroform (Feist and Hadi 1965; Reif-Acherman 2009) and the concomitant removal of heat from a surrounding area. These 'static refrigerators' are not used at present: the current evolution (dynamic refrigerators) concerns the continuous circulation of cold air between different areas; moreover, frost formation is avoided by means of forced ventilation and thermal changes caused by frequent door openings are remarkably reduced.

In the industrial field, at the end of the preparation or the last heat treatment, foods or intermediate products should reach refrigeration temperatures very quickly because microbial spreading is favoured between 3 and 70 °C. Consequently, 'blast chillers'—discontinuous (cabinets) or continuous (tunnel) systems—are frequently used: refrigeration conditions are reached by means of the introduction of cold air (forced ventilation at about 1 °C) or cold water. Subsequently, treated foods have to be stored in 'permanent' chillers and/or dedicated trucks or rail cars capable of self-cooling. By the legal viewpoint, the transportation has to be carried out with the absolute warranty (digital or printed records) of thermal values below or equal to −18 °C (only for frozen products and long distances). Otherwise, simple small trucks with thermally insulated walls are required.

2.2.1.2 Freezing Techniques

Basically, freezing involves the exposure of foods and beverages to low or very low temperatures with the consequent water crystallisation and the solidification of the whole product. A food is defined 'frozen' when 80–90 % of the total water amount is turned into ice.

It has to be noted that freezing methods affect only free water; a small fraction of aqueous amounts (2–5 %) is electrostatically bound to sugars, proteins and solutes; this water is also named 'non-freezing water' (Ali and Bettelheim 1985). Consequently, foods and beverages—liquid, solid or fluid water solutions—have to reach temperatures below 0 °C (the cryoscopic point or frost point of pure water at atmospheric pressure) for freezing purposes. Generally, the freezing point of free water is initially between −4 and −0.5 °C, but the freezing treatment should obtain a 'core' temperature between −10 and −25 °C (end of the process). In fact, a notable fraction of free water is constantly separated from the original food solution in form of ice. The end of the process is reached when 80–90 % of the total water content is turned into ice: the remaining non-freezing solvent and a certain amount of free water are also present. Interestingly, the residual free water (with a notable concentration of dissolved molecules) is located in very small channels and may be also able of free movement through the treated food.

Basically, the freezing process may be subdivided into two steps:

(a) Nucleation and crystallisation. Temperature values: between 0 and −7 °C. Ice crystals are initially formed; this step implies the maximum separation of water in the solid state from the remaining food system, including non-freezing and residual free water

(b) Crystal growth. Each ice crystal tends to augment its dimensions with the incorporation of other water molecules around the initial nucleus.

It has to be noted that the dimension of final ice crystals is inversely proportional to the number of initial ice nuclei. Moreover, the number of ice crystals depends on the freezing speed. Basically, two different situations can be observed (Zhao 2007):

(1) Slow freezing. The food is subjected to temperature values between −8 and −20 °C with a low penetration speed. Should this be the situation, the initial number of ice crystals would be reduced (poor nucleation and crystallisation). Consequently, strong crystals of considerable size are easily observed. Unfortunately, the formation of a reduced number of high-scale ice crystals is the main cause for micro- and macro-damages in the food because macro-crystals easily damage cell walls (vegetable and animal tissues). Once defrosted, the food does not return to the original texture and fibrous-like products can be obtained; additionally, unpleasant flavours are often reported. For these reasons, the method is not recommended. Recently, new similar systems have been proposed: high-pressure freezing and dehydrofreezing (initial dehydration of foods and subsequent freezing), with the aim of avoiding volumetric augments and tissue damages (Li and Sun 2002)

(2) Rapid freezing (required systems: air freezers; plate freezers; liquid-immersion freezers). The food is subjected to temperatures between −30 and −50 °C with notable penetration speed. Should this be the situation, the initial number of ice crystals would be remarkable (good or excellent nucleation and

crystallisation). Consequently, many microcrystals are easily observed (a few μm only). The formation of a notable number of small-scale ice crystals can be very useful because microcrystals do not easily damage cell walls (vegetable and animal tissues). Once defrosted, the food may return to the original texture with potentially small variations: fibrous-like products are not reported when speaking of rapid freezing. Sensorial features are substantially preserved: higher shelf life values have been also demonstrated. For these reasons, the method is recommended in the industrial ambit. The so-called 'individual quick freezing' (IQF) system is extremely useful when foods are not packaged. However, the main problem remains the superficial dehydration with related weight loss between 2 and 3 % (Pruthi 1999; Senesi 1984)

(3) Flash or cryogenic freezing. The food is subjected to temperatures below −50 °C with high penetration speed. Should this situation be observed, the initial number of ice crystals would be higher than expected in comparison with rapid freezing. Consequently, the performance of this method is reported to be excellent in comparison with rapid freezing, despite some observation with concern to the real necessity (Farouk et al. 2004). The total treatment time is reduced (maximum time: some hours, depending on the dimensional features and composition of products). Once defrosted, the food may return to the original texture with potentially small variations: fibrous-like products are not reported when speaking of rapid freezing. Sensorial features are substantially preserved: higher shelf life values have been also demonstrated. For these reasons, this method is recommended in the industrial ambit. Generally, the use of liquid nitrogen or solid carbon dioxide is required in flash (cryogenic) systems because products are usually placed on a conveyor belt into dedicated tunnels and subsequently cooled with sprayed or dribbled cold gas. Subsequently, the product must reach the final temperature. Dehydration is substantially negligible

(4) Deep freezing. This technique is discussed in Sect. 2.2.1.3.

The duration of treatments can vary, depending on the volume and thickness of the product, from a few minutes to a few hours.

2.2.1.3 Deep Freezing

Deep-frozen foods are products subjected to a special freezing process: the zone of maximum crystallisation has to be reached quickly and maintained constant in the whole product at temperatures ≤ -18 °C until the end of durability periods. The mentioned thermal value is compulsory from the legal viewpoint. This process, also named 'ultrafast' system, can be performed with air or plate freezers: the only exception concerns the immersion in non-freezing liquids. Requisites are:

- The inner temperature (at the centre of the whole product) has to be minor or equal than −18 °C
- The product has to be continually kept at temperatures ≤ −18 °C until the sale to the final consumer.

Moreover, stricter conditions may be applied in several stages because of inevitable thermal changes in some point of the whole food chain. In particular, factory warehouses may be kept at −30 °C before the delivery of frozen foods, while intermediate storage chillers (in all possible points of the food chain except for the initial production plant) may remain at −25 °C value (subcooling).

It should be also noted that:

(1)   The use of peculiar packaging materials is strictly required when speaking of deep-frozen foods. These food packaging materials include aluminium (alone or combined with plastic films), tinplate, cardboard, coupled plastic films (nylon/polyethylene, polypropylene/polyethylene)

(2)   The label must contain all required information for the proper use of the food, in accordance with official Regulations.

2.3   Non-thermal Food Preservation Methods

Food products can be preserved by means of thermally based preservation methods or different (non-thermally conducted) systems. Actually, this subdivision is not so clear because a peculiar food product can be subjected to two different preservation systems. Basically, the main difference between above-mentioned procedures is the definition of 'thermal processing method'. It may be inferred that all processes implying the use of heat (or heat removal) can be defined 'thermal' procedures.

Sterilisation procedures concern only preserved foods, which are packaged in a hermetically sealed container and have undergone a specific heat treatment or other authorised treatments with the aim of inactivating irreversibly dangerous enzymes and microorganisms. Two basic categories of sterilised foods (Sect. 2.1.2, Table 2.1) are known: acidic products with pH ≤ 4.5 and non-acidic foods with pH ≥ 4.5 (Soncini 2008).

On the other hand, semi-preserved foods are products whose stability over time depends on the preservation method and environmental conditions. In detail, the following products can be defined semi-preserved foods:

(a)   Pasteurised foods with pH ≥ 4.5. These products are packaged in a hermetically sealed container, pasteurised (or subjected to analogous treatments with lower performances if compared with sterilisation) and their shelf life depends on environmental conditions (low temperatures, use of modified atmosphere, etc.). For these reasons, pasteurised semi-preserved foods have to be refrigerated

(b) Non-thermally treated semi-preserved foods. These products are packaged in more or less hermetic containers and have undergone a specific technological treatment without heat addition. In addition, their performance and safety depend on storage conditions. Allowed technological systems are: smoking, drying, lyophilisation and addition of natural substances with inhibiting power. These products are normally stored as refrigerated or frozen foods.

The main objective of this chapter is the comparison of different preservation systems and techniques. Because of the existence of different foods and beverages belonging to the category of preserved (sterilised) products and the more complex world of semi-preserved foods, a critical comparison of different systems should be carried out. In addition, the association between several peculiar methods of technological preservation and selected thermal methods should be explained with reliable argumentation. Next sections are dedicated to a brief overview of non-thermal preservation systems (Fig. 2.2) and the explanation of synergic effects between these methods and thermal procedures.

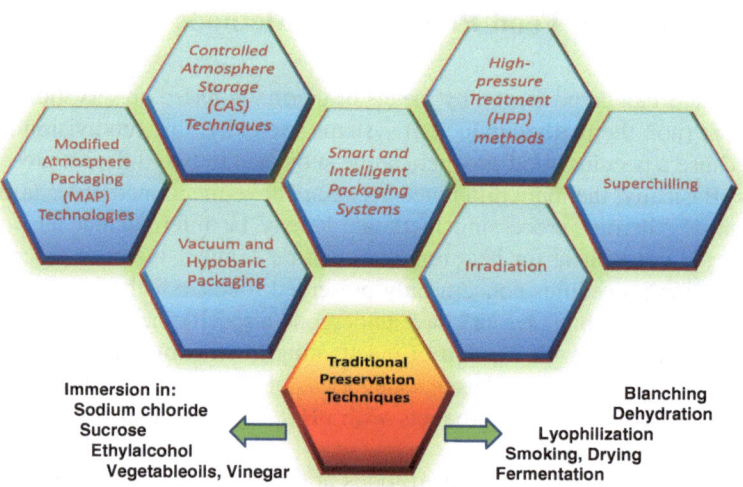

Non-Thermal Food Preservation Methods

Fig. 2.2 A brief overview of non-thermal preservation techniques. The main difference concerns the use of traditional preservation systems such as the immersion in salt solutions, sucrose, ethyl alcohol, vegetable oils and vinegar, or the use of 'historical' methods such as smoking, drying and fermentative processes. Actually, other systems may be considered in this category with the aim of discriminating them in comparison with high-technology procedures: dehydration, lyophilisation and blanching. High-technology preservation systems—MAP, CAS, intelligent packaging, superchilling, etc.—are often used in modern plants and distribution services

2.3.1 Modified Atmosphere Packaging Technologies

2.3.1.1 MAP Systems: General Features

Modified atmosphere packaging (MAP) techniques are one of the most recent and interesting applications in the field of food preservation. Basically, this technology can enhance shelf life values for certain products and reduce the addition of food preservatives.

Briefly, food degradation caused by chemical–physical reactions and microbiological spreading (enzymatic reactions are also included) can be delayed or inhibited to a certain extent by means of the simple modification of atmosphere compositions in the final food packaging (Micali et al. 2009; Parisi 2009, 2012). In this ambit, two possible MAP technologies can be used for food applications:

(1) 'Active' MAP systems. The existing gas mixture into the container is removed and a different gaseous composition is inserted before packaging
(2) 'Passive' MAP methods. The composition of food packaging materials is designed to allow the continuous modification of inner atmospheres caused by cellular 'respiration'. This phenomenon is observed when speaking of fresh (minimally processed) fruits and vegetables, with special attention to cut products. Cellular respiration causes the continuous increase of carbon dioxide at the expense of oxygen; consequently, exceeding carbon dioxide has to be removed (Senesi 1984).

The effectiveness of MAP systems depends on the 'empty' volume occupied by the gaseous mixture (it should be equal to the space occupied by the packaged food). In addition:

(a) The composition of modified atmospheres depends on the peculiar food. Generally, most used gases (in different proportions) are carbon dioxide, oxygen, nitrogen and/or argon; in particular, carbon dioxide may be extremely useful (Molin 2000). Possible mixtures have to be designed on the basis of the microbial ecology and approximate chemical compositions of packaged foods (Ercolini et al. 2006a). Some specific parameters such as water activity (a_w), pH and redox potential have to be taken into account; the latter of these measures, in particular, can be very useful to try to determine the probability of growth and survival of pathogenic and spoilage microorganisms (Delia et al. 2005; Parisi et al. 2004; Senesi et al. 2000)
(b) The composition of packaged foods has to be taken into account, specifically with relation to fatty and moist foods and possible irregular agglomerations into MAP products (Sect. 2.3.1.2)
(c) The augment of food contact surface is critical. As an example, diced cheeses (Sect. 2.3.1.2) can show enhanced shelf life values if compared with whole cheeses in the same MAP condition because the food/atmosphere interface is dimensionally increased. This interesting result is not expected normally

because cuts or different mechanical operations aiming to separate or subdivide solid foods cause generally lower durability values (Delia et al. 2005; Parisi 2002, 2009; Parisi et al. 2004).

2.3.1.2 MAP Systems: Technological Details

The 'empty' volume occupied by the gaseous mixture in MAP products has a remarkable importance. This space should be equal to the space occupied by the packaged food. However, this condition is not always possible because of the physical disposition of the edible mass into the package. In other terms, subdivided food products can physically occupy part of the theoretical empty volume into the MAP container. This situation has been reported clearly for MAP diced mozzarella cheeses (Parisi et al. 2004). This product tends to occupy large volumes of the inner package because of the irregular disposition of cheese cubes and the fatty composition of products, with consequent agglomeration and partial 'fusion' between single cheese parts.

MAP technology requires the combination of three interdependent elements: the packaging machine, the type of wrapping material and the composition of gaseous mixtures. In detail, the release of modified atmospheres (after the removal of pre-existing gases into the container) can be observed immediately before, during or immediately after the placement of whole or subdivided foods. With relation to gaseous mixtures, the so-called 'reactivity' or 'inertness' depends on the abundance of three main gases: carbon dioxide, nitrogen and oxygen (Micali et al. 2009). Should nitrogen be mainly present, the inner gas mixture would be defined 'inert'; on the contrary, mixtures with carbon dioxide and/or oxygen (with the exclusion of inert nitrogen) are defined 'reactive'. Intermediate compositions (carbon dioxide and nitrogen, with or without oxygen) are considered mixtures with a limited 'reactivity' by the technological viewpoint (Micali et al. 2009).

Substantially, it can be affirmed that MAP technology offers excellent results in terms of food durability, if compared with known traditional techniques such as vacuum packaging solutions. In addition, sensorial features remain acceptable. As a result, many different foods can be preserved in this way: baked products, pre-cooked foods, coffee, fruit juices, wines, chips, processed meats, etc.

MAP advantages are certainly referred to the commercial stability and the safety of products; at the same time, long shelf life values allow the optimisation of logistic operations and a more efficient management of transportation costs without the use of declared preservatives. Moreover, the 'thorny' matter of waste foods (expired products) may be partially solved.

However, the MAP technology cannot assure the complete safety and hygiene of treated food products without a solid and reliable basis: the use of good or excellent raw materials and packaging materials (and objects). In other words, the basic 'support' for MAP systems (and each preservation technique) is always the initial raw material (or the sum of used raw materials). As above-mentioned, some chemical–physical data such as pH, a_W and redox potential have to be taken into account. In detail, redox potentials can be extremely useful because of the direct connection between these values and the probability of microbial spreading for selected microorganisms.

The determination of redox potentials in foods is recommended because of the connection between the increase towards higher values and the concomitant decline in the multiplication of life forms such as *Listeria monocytogenes*, *Escherichia coli* and *Salmonella* spp. Because of the influence of the edible 'culture medium' (the food itself) on the redox behaviour, several foods are not good enough when speaking of microbial spreading; as a consequence, these foods could be treated with MAP technologies with the aim of maintaining initial redox potentials in an unfavourable way for many anaerobic bacteria (Delia et al. 2005).

Generally, the first objective of MAP systems is the reduction of contacts between oxygen and food surfaces. In fact, oxygen is mainly responsible for the promotion of microbial spreading (aerobic or facultative anaerobic life forms). On the other hand, this gas could be useful in certain situations: as an example, the bright red colour of oxyhemoglobin in red meats and fish products can be notably enhanced in presence of oxygen; otherwise, the formation of meta-myoglobin would be easily expected with inacceptable modifications of colorimetric performances (Delia et al. 2005). However, oxygen can favour lipoxidation and fat rancidity in medium-fat and fat fish such as salmon and tuna fish. In addition, food degradation may be ascribed to the oxidant action of this gas, including the degradation of β-carotene.

On the other side, a real antimicrobial action is ascribed to carbon dioxide (this effect is maximum when the related concentration exceeds 30–40 % of the total gaseous mixture). Naturally, the efficacy of carbon dioxide depends also on the microbiological state of treated foods, in terms of initial microbial counts. On the other side, high concentrations of this gas may lead to fractures in the packaging material (collapse) with consequent severe alterations of sensorial features. The explanation of this strange phenomenon is mainly ascribed to the absorption of hydrated carbon dioxide (carbonic acid) in foods, with the concomitant pH lowering and the decrease of inner pressures into packages (Delia et al. 2005). pH low values are efficient enough in terms of microbial inhibition; in addition, the production of biogenic amines is delayed (Micali et al. 2009). The effect of this gas on vegetable products is substantially the inhibition of cellular respiration; moreover, maturation processes are delayed and possible damages to vegetable tissues because of cold temperatures are notably reduced (Micali et al. 2009). With reference to negative effects, it should also be remembered that excessive amounts of carbon dioxide may cause dehydration in meats.

Nitrogen is completely inert: as a consequence, there are no possible interaction with organic substances. Basically, this gas is used as a filler with the aim of preventing possible packaging collapses, if carbon dioxide is used and possibly dissolved in foods. Because of the complete inertness, pH is not affected.

Other possible gases are:

(a) Carbon monoxide. This toxic gas has not inhibitory effect on microorganisms. However, its use has been often reported because of positive sensorial effects such as the formation of carboxymyoglobin with red colour enhancement in certain products (beef, fish). On the other hand, carbon monoxide is questioned because of possible effects on food operators. In addition, the above-mentioned formation of red-coloured carboxymyoglobin may be used for fraud purposes

(b) Argon and other noble gases. These chemical elements may be a good alternative because of the substantial inertness, colourless, odourless and tasteless. In addition, gases such as argon can prevent crushing and deformation damages in MAP products.

2.3.1.3 MAP Systems: Synergic Applications with Thermal Processes and Possible Failures

The use of MAP systems can be very useful if these technologies are compared to more traditional (air or vacuum packaging) techniques. At present, many applications are possible when speaking of pasteurised and cooked foods: meat and sausages, baked foods, precooked foods, cheeses, milk powders, coffee, fruit juice, wine, chips, pretzels, etc. The additional removal of food preservatives has to be highlighted.

On the other side, possible failures of MAP systems can be caused by:

- Initial conditions of foods and beverages. In other words, incorrectly stored, manipulated or pasteurised foods cannot be well preserved with MAP (or vacuum) technologies. In particular, the possibility of certain greenish colours (Kröckel 1995) on the surface of processed meats may be favoured when using peculiar MAP gases
- Strange behaviours of packaged foods. In detail, certain sensorial variations have been often observed and correlated to the dissolution of MAP gases (carbon dioxide) in foods, where possible (cheeses)
- Inner or external package fractures caused by the permeability of food packaging materials, with consequent lower pressure into the MAP product (Parisi 2012; Sivertsvik et al. 2002). Other situations depend on pre-existing food packaging failures (Parisi 2013) or processing errors (during the replacement of the inner atmosphere)
- Incorrect storage temperatures (these products have to be stored in refrigerated chillers at least).

The above-discussed failures can be also observed in vacuum packaged foods: naturally, the frequency of possible defects, including food safety and hygiene situations, depends on the peculiar preservation system. In other terms, MAP methods aim at modifying shelf life values of foods by means of the controlled variation of inner atmosphere; on the contrary, vacuum packaging techniques aim at inhibiting the development of microbial spreading and chemical–physical reactions based on the necessary presence of oxygen.

2.3.2 Controlled Atmosphere Storage Techniques

2.3.2.1 CAS Systems: General Features

Because of the effect of oxygen on processed and fresh foods (Sect. 2.3.2), another preservation system—controlled atmosphere storage (CAS)—can be considered into large-scale food warehouses in synergy with refrigeration procedures. In brief, the air inside refrigerated cells can be ejected and replaced with mixed gases, depending on the type of food product. The composition of 'artificial' gas mixtures is constantly monitored and kept constant by means of automatic control systems (analysers, absorbers and gas generators). Generally, atmospheres should have the following rough composition: nitrogen (92–95 %); carbon dioxide (2–4 %); oxygen (3–4 %); the temperature of chillers should not exceed 3–4 °C. Basically, the main modification concerns the substitution of oxygen with nitrogen, carbon dioxide or their mixtures (Sect. 2.3.1). CAS systems are mainly used in food warehouses and sea freight containers for apples, pears, citrus fruits, other fruits and vegetables, and flowers. Stored products are generally sensible to alterations because of cellular respiration phenomena and the consequent production of carbon dioxide, water vapour and several organic and aromatic compounds.

2.3.2.2 CAS Systems: Synergic Applications with Thermal Processes and Possible Failures

The CAS system may avoid the early degradation of fresh products if combined with cold-based technologies. In detail, the production of ethylene is reduced; the degradation of chlorophylls may be blocked; pectins may be preserved at some extent from hydrolysis. Stored products can be subsequently used for different productions and other preservation treatments, including thermal processing. With reference to possible defects, observed failures are generally correlated to the incorrect management of process parameters: chiller temperatures, adequate ventilation, surveillance and the continuous modification of inner atmospheres into chillers.

2.3.3 Vacuum Packaging and Hypobaric Packaging Systems

2.3.3.1 Vacuum and Hypobaric Packaging Systems: General Features

The amount of oxygen inside food packages may be also reduced if the inner pressure is lower than 2.66×10^3 Pa or between this value and 101,300 Pa (atmospheric pressure). Substantially, two different methods may be used in this ambit:

(1) Vacuum packaging. The inner pressure is lower than 2.66×10^3 Pa. Recommended containers for these methods can be pouches, special rigid containers and thermoformed trays. The virtual absence of oxygen does not favour the aerobic activity of many microorganisms; consequently, deterioration processes are remarkably delayed, and sensorial features (appearance, colour, aroma, flavour and texture) suffer some little modification. The same thing can be affirmed for chemical and nutritional features (digestibility, protein amount, vitamins and starches). Moreover, environmental contaminants are removed with the originally contained air into the package
(2) Hypobaric packaging. The inner pressure is between this value and 101,300 Pa. Performances can be interesting enough in terms of shelf life values, sensorial expectations and safety, although vacuum packaging is considered with more favour.

2.3.3.2 Vacuum and Hypobaric Packaging Systems: Synergic Applications with Thermal Processes and Possible Failures

Vacuum and hypobaric methods can be used in association with low temperatures (refrigeration) and thermal processes in general (including the simple steam cooking at 80 °C, for catering purposes) with excellent performances.

In particular, vacuum packaging meets fundamental food safety requirements. The main advantage is the extension of shelf life values: products are kept longer under refrigeration conditions and during the commercial transportation. Moreover, the possible addition of chemical preservatives, salt or strong spices can be advantageously replaced or reduced. Another important advantage, from the economic angle, is correlated with the virtual absence of weight loss because of evaporation and liquid release.

On the other side, vacuum techniques may be correlated with the possible modification of aroma and the limited evaporation of moisture and volatile compounds. However, the main failure for vacuum and hypobaric packaged foods is always correlated with the use of food packaging materials in terms of possible damages before use, incorrect design/choice and poor 'barrier' properties (Parisi 2012).

2.3.4   Smart and Intelligent Packaging Systems

2.3.4.1   Smart and Intelligent Packaging: General Features

The increasing demand for fresh products and the consequent necessity of reliable demonstrations—including also new and redeveloped analytical techniques (Parisi 2016)—has progressively stimulated the development of the so-called 'functional' packaging. Generally, these packaging materials are subdivided in two different categories: 'active' and 'intelligent' packaging systems (Parisi 2009), depending on the related function.

Briefly, 'active' or 'smart' packaging materials are designed with a very specific aim: the continuous (active) modification of food and atmosphere properties during the whole shelf life period (Parisi 2009) with the consequent enhancement of positive food properties, including safety. This objective can be achieved by means of the chemical interaction with the packaged food and the gas mixture inside the package. Smart systems can allow the possibility of an active control because of the direct and predictable interaction with packaged foods. As a result, food technologists may be able to influence the complex phenomena of chemical, microbiological, physical and mechanical modifications occurring during storage. Finally, it should be remembered that modern packaging materials have also a clear marketing function (Brunazzi et al. 2014); consequently, the use of smart systems should be evaluated in relation to preservation purposes and marketing strategies (Parisi 2012).

On the other side, 'intelligent' packages are designed with the aim of reporting product changes during the whole shelf life period, with the consequent possibility of diminished freshness and/or durability. The expression 'intelligent packaging' means a packaging technique that involves the use of an internal or external indicator, capable of actively representing the history of the product and the remaining shelf life (Parisi 2009).

The correlation of smart and intelligent systems with thermal processing is not apparently clear. Both solutions require the use of other preservation strategies, including thermal processing systems for unpackaged foods and cold-based techniques (refrigeration). Thermally processed foods can be packaged with modified atmospheres and interact with active compounds such as antimicrobials, antioxidants or other useful substances. The continuous modification of food composition can also be obtained by means of the variation of gaseous mixtures (head-space control): small accessories containing gas absorbers or scavengers (target gases: oxygen, ethylene, moisture) or generators of oxygen, carbon dioxide or ethanol can be used (Parisi 2009). In detail, the following systems are currently reported:

- Moisture absorbers (glycerol, silica gel, clays) for meats
- Humidity regulators (sodium and potassium chloride)
- Ethylene absorbers (activated carbon, zeolites) or emitters of carbon dioxide (ascorbic acid) for vegetables

- Carbon dioxide absorbers for packed coffee (calcium chloride and sodium or potassium hydroxide)
- Ethanol emitters for baked goods.

Moreover, the active packaging may induce natural antimicrobials such as essential oils or bacteriocins: these substances can inhibit microbial growth in foods (Ercolini et al. 2006b; Mauriello et al. 2004; Murray and Richard 1997; Ouwehand 1998; Skandamis and Nychas 2001; Tassou et al. 2000). However, strict management requirements are needed, and storage procedures (good manufacturing practices, high hygienic standards) are also required.

Anyway, active systems cannot fully give expected performances without reliable traditional preservation techniques. Cold storage is required: in fact, the design and the implementation of active approaches are based on the preventive application of refrigeration systems. Sometimes, these applications have not been recommended when speaking of frozen foods.

The intelligent packaging has to interact with the environment and the whole food/packaging system: recorded variations such as temperature abuses should be immediately visible by means of evident chromatic variations. Time temperature indicators (TTI) represent the most widespread forms of commercially available intelligent packaging. TTI are normally applied on the outer surface of packages: the thermal history of the whole food/packaging system is recorded and made available readily (Parisi 2009). In addition, some commercial application of oxygen and carbon dioxide indicators already exists: these applications—tablets or strips into containers—can vary their colour because of the effect of specific chemical and enzymatic reactions on the composition of inner atmospheres. Otherwise, a destructive analytical procedure would be required.

2.3.4.2 Smart and Intelligent Packaging: Synergic Applications with Thermal Processes and Possible Failures

Because of the peculiar design, smart and intelligent systems should be used if coupled with cold-based storage methods. For this basic reason, related advantages should be considered simply in terms of performance enhancement for cold-stored foods or refrigerated products packed with modified atmospheres. On the other hand, failures can be easily defined as the opposite of predictable (desired) results:

(a) Insufficient modification of food properties
(b) Insufficient modification of atmosphere properties
(c) Inaccurate demonstration of thermal history
(d) Inaccurate demonstration of qualitative variations (with relation to foods and/or gas mixtures).

Moreover, each mechanical damage caused by one or more of the above-mentioned defects can be ascribed (at first sight) to different reasons with the exclusion of smart devices. Actually, the simple insufficient modification of gaseous compositions into packages can easily cause packaging defects such as fractures or container collapsing (Parisi 2012).

2.3.5 High-Pressure Treatment

An interesting non-thermal preservation process is based on the application of high pressures (Hugas et al. 2002; Patterson 2005; Simpson et al. 2012). This technique, generally named high-pressure processing (HPP), shows very good performances when speaking of food preservation because of the effect on several spoilage bacteria and pathogens (destruction) and some enzymes (inactivation). HHP may significantly reduce the presence of *Salmonella* spp and *L. monocytogenes* in raw meats and marinades.

HPP macroscopic effects on processed foods are different (pressures >200 MPa with maximum values >500 MPa):

(1) Inactivation and death of microorganisms
(2) Structural and functional modification of biopolymers (proteins, enzymes and polysaccharides). In particular, the tridimensional structure of enzymes protein molecules suffers important variations (Simpson et al. 2012)
(3) Morphological changes (vacuolar gas compression, deformation/elongation)
(4) Altered metabolism and biochemical reactions
(5) Changes in the cell membrane (altered permeability).

The use of high pressures (400–700 MPa) on foods causes the permeabilisation of cell membranes and other phenomena such as the probable crystallisation of membrane phospholipids. Anyway, the microbial resistance to high pressures depends on several factors: the type of microorganism, its stage of development, process conditions [time, pressure, temperature, chemical parameters of treated foods (a_w, pH, etc.)].

HHP is recommended for the microbial stabilisation of acid and heat-sensible foods, the sanitisation of non-acidic products and the change in the technological functionality of ingredients. In addition, solid and liquid foods can be treated with HPP techniques; post-contamination is minimised; cleaning and sanitising procedures in food plants are easier.

With relation to potential resistances, it has to be highlighted that viruses and spores are very resistant to HHP; on the other side, HHP shows excellent performances with non-sporeforming pathogens (*L. monocytogenes, Staphylococcus aureus*, etc.).

Other potential disadvantages of the HPP system for packaged foods concern substantially the complex manipulation of foods, the limited choice for containers, long cycle times and the expensive cost of HPP lines. Actually, non-packaged foods can be HPP treated with interesting advantages: easy handling, possibility of many different containers and the absence of loading and unloading times; the process may be defined 'continuous' (Parisi 2005). On the other side, only liquid and semi-liquid foods can be considered; in addition, aseptic packaging systems are required.

2.3.6 Irradiation

The so-called 'cold pasteurisation' has certainly interesting effects on enzymatic reactions and microbial activity. Basically, the exposure of foods to ionising radiations (X-rays, accelerated electron beams and γ-rays) enhances the shelf life expectation of certain products such as vegetables, fruits, meats and seafoods (Simpson et al. 2012). On the other hand, consumers appear to reject irradiation with relation to non-food accidents (Donati 2015); moreover, some decrease in nutrients has been reported (Simpson et al. 2012) and the use of this technique can be effective on condition that enzymatic inactivation is complete. For these reasons, the method is certainly powerful, but the current market does not seem to accept it at present.

2.3.7 Superchilling

Superchilling is an interesting method for the preservation of ced white fish on ships (Waterman and Taylor 2001). Substantially, the process concerns the partial freezing of water inside foods: the final temperature should be 1–2 °C below the initial freezing point. A notable extension of shelf life has been observed (Kaale et al. 2011; Magnussen et al. 2008; Schubring 2009), thereby extending the storage life of the fish. However, three main disadvantages have to be considered (Waterman and Taylor 2001):

(1) Superchilled products cannot be processed or used immediately because they are partially frozen. A certain time loss is expected
(2) Normal quality inspections cannot be performed on partially frozen fish. Once more, a certain time loss should be taken into account
(3) Superchilling parameters (temperature) have to be strictly monitored and constant. This condition could not be always observed easily on trawlers.

2.4 Traditional Preservation Techniques

Basically, 'traditional' preservation methods involve the use of food additives with microbial actions (natural or artificial origin), peculiar historical systems such as smoking and fermentation, and the addition of natural food ingredients. This section is dedicated to a brief description of these techniques, with possible advantages and risks if compared with thermal preservation.

2.4.1 Preservation with Traditional Additives and Pretreatments

2.4.1.1 Preservation with Sodium Chloride

The traditional use of salt for the preservation of meat, fish and vegetable products (Delgado et al. 2016) can also provide particular sensorial features. With exclusive relation to sodium chloride, the following positive actions have to be considered:

(a) Increased osmotic pressure and consequent damages to microbial cells. Osmotic pressure depends exclusively on the number of dissolved solute particles in solution, while their size is not important. Consequently, small-size ions such as sodium cations and chloride anions can provide higher osmotic pressure values (and a better preservation of treated foods) if compared with other high-size solutes such as sucrose at the same concentration (g/l)

(b) Decrease of a_w values with the consequent inhibition of microbial activities. Sodium and chloride ions tend to 'block' the surrounding water molecules: the amount of bioavailable free water has to decrease necessarily. In addition, the formation of hydroxide ions reduces a_w. Finally, sodium chloride and proteolytic enzymes may create new compounds with the consequent reduction of cellular decomposition.

The action of the salt is mainly bacteriostatic; however, the bactericidal action cannot be guaranteed because of the possible presence of osmophilic, halophilic or salt-tolerant life forms. For this reason, salting is often combined with other preservation methods such as low temperatures, the addition of nitrite and nitrate, cooking or smoking process.

Usually, commercially available salt is obtained from the evaporation of seawater, from natural brines and from underground warehouses. Different methods are available at present, including the addition of salt in direct contact (possibly mixed with spices, nitrates or nitrites), or by rubbing, or with stratification in alternate layers. Anyway, salt should be added at low temperatures: otherwise, certain defects could occur including dishomogeneity of water transfer to the external surface of foods and the formation of superficial crusts. Other well-known systems comprehend the addition of salt by means of the brine injection into meat tissues or by simple immersion (e.g. Grana and Parmigiano Reggiano cheeses). Anyway, cooking and smoking processes can be combined with salting.

2.4.1.2 Preservation with Sucrose

Sucrose can increase the osmotic pressure similarly to salt, but its preservative action is low if compared with sodium chloride; for this reason, high concentrations of added sucrose are needed (65–70 %). Consequently, treated foods are often jams, fruit jellies and candies.

It should be also noted that higher concentrations of added sucrose lead to crystallisation phenomena, while lower quantities cannot be effective enough (a synergic action with other preservation methods such as concentration, acidification and cold-temperature storage are required). Glucose and fructose may be used as possible sucrose substitutes because of their lower molecular weight and the consequent reduced amount (40 %).

Similarly to salt, sucrose has a prevailing bacteriostatic action without bactericidal effects; moreover, osmophilic life forms may remain unaffected in presence of sucrose. For these reasons, a synergic action with other methods is needed and recommended.

2.4.1.3 Preservation with Ethyl Alcohol

Ethanol—a non-toxic alcohol for food applications—is added to fruits for the production of peculiar products with amounts between 50 and 70 %. In these conditions, ethanol would be lethal against vegetative forms; naturally, treated foods would also obtain a pleasant and strong taste. On the other hand, initially present vitamins would be destroyed.

2.4.1.4 Preservation with Oils

Preserved foods in oil are one of the most 'popular' products at present (Delgado et al. 2016). Generally, two main food products—vegetables and fish—are preserved with vegetable oils. Intermediate foods are placed in dedicated containers (after cooking or other thermal treatments) and immersed in oil with the aim of covering completely the edible product. Substantially, the aim of food technologists is to create an oxygen-free environment, with the consequent inhibition of aerobic bacteria; vegetable oils do not show a direct preservative action.

For this reason, the use of oils is not effective and recommended when speaking of possible anaerobic activity by certain microorganisms (example: *Clostridium botulinum*). Moreover, moist foods cannot be immersed in oil and remain preserved because living microorganisms may be able to spread in the aqueous phase (example: undrained foods). Consequently, this traditional method may require the synergic action of different preservation systems such as pasteurisation and sterilisation (canned tuna is initially cooked before being salted, canned in oil and sterilised) or the addition of salt and vinegar (artichokes are boiled in water and vinegar before treatment in oil).

The storage in oil may determine a certain loss of nutrients (vitamins above all); in addition, the preservation of pickles is very different if compared to fresh products. Moreover, the composition of drained foods gives an excess of energy in comparison with the untreated product: sodium contents are rather high and lipids may be still 10 times greater if compared with fresh products.

2.4.1.5 Preservation with Vinegar

Vinegar is historically used to preserve vegetables and some fish products by means of its content in acetic acid (minimum quantity: 6 %). Substantially, the immersion in vinegar can determine:

(1) A microbiostatic action (pH values are notably lowered)
(2) A bactericidal action because of the known toxic effect of acetic acid.

The immersion in cold or heated vinegar (75–80 °C) has to be carried out after preliminary treatments: foods should be seared, salted or macerated in vinegar for a few days. Moreover, the acid solution can be 'enriched' with spices or herbs. These procedures allow to counteract the possible decrease of preservation effects because of the excessive dilution (intermediate solutions contain always more or less moist foods).

Almost all vegetables (peppers, eggplant, cucumbers, capers, etc.) can be preserved in vinegar. Initial foods can be either fresh or semi-preserved raw materials; in the latter case, these are kept in brine up to the time of the preservation process, when they are rinsed and put under vinegar.

The modern production of preserved foods in vinegar involves also the use of chemical additives (citric acid and ascorbic acid) or a pasteurisation treatment. For these reasons, radical modifications of organoleptic features (taste, colour and odour) have to be expected. Moreover, vitamins and other nutrients are likely lost. On the other time, pickles improve with time and can be stored for a very long time.

2.4.1.6 Blanching

Blanching is a thermal pretreatment with peculiar applications for vegetable raw materials. Actually, the process should be discussed in Sect. 2.1 because of the use of heat (water or steam at 70–105 °C). However, the importance of blanching is marginal when speaking of food preservation methods.

In fact, the main purpose of this technique is to inhibit only enzymatic activity in vegetables and some fruit types before subjecting them to 'classical' preservation systems. In other words, blanching has a limited importance when speaking of food safety, inhibition of spoilage microorganisms and elimination of pathogen agents; on the contrary, further preservation methods are required. By the chemical viewpoint, macroscopic effects of blanching are the modification of sensorial features (colours above all) because of the transformation of chlorophylls and other organic substances such as β-carotene, ascorbic acid, etc. (Muftugil 1986; Negi and Roy 2000). Other effects include the elimination of unpleasant odours and gases enclosed into plant tissues (with reduced oxidation risks), the decrease of microbial counts and the coagulation of proteins with lower cooking times.

From the technological viewpoint, raw materials are rapidly heated to the desired temperature: after a certain time, they are rapidly cooled to room temperature.

2.4.2 Other Traditional and Modern Preservation Techniques

2.4.2.1 Dehydration

The term 'dehydration' means the progressive reduction of water content inside foods with the aim of inhibiting microbial spoilage; essentially, this process has microbiostatic effects. However, three different processes can be named 'dehydration':

(a) Concentration
(b) Drying
(c) Freeze-drying.

Basically, these methods are performed with the aim of reducing the bioavailability of free water into foods. This availability is measured in terms of water activity; should a_w (Sects. 2.3.1 and 2.4.1.1) be low enough, the decrease in the growth rate of living microorganisms would be observed until the complete block of all metabolic activities. a_w can be reduced:

(1) By means of the removal of water molecules from foods (dehydration), or
(2) By means of the dissolution of peculiar substances (solutes) in the solid food solution (chemical preservation).

Dehydration has a certain microbiostatic action; in addition, water bioavailability is remarkably reduced. On the other side, toxins and spores remain substantially unaffected; for this reason, the use of dehydration cannot be recommended when using raw materials with appreciable microbial counts.

In general, the tripartite subdivision of dehydration systems depends on the technological solution for the elimination (separation) of the aqueous solvent from the food matrix.

Concentration techniques are generally performed on liquid foods or products with high water amounts; the objective is the reduction of volumes and weights (limitation of transportation costs). It has to be noted that preservation effects are rather limited and always associated with other treatments (pasteurisation, freezing, addition of chemicals, etc.).

Concentration by evaporation is the most used technique for the production of concentrated feed and food products. A basic requirement has to be satisfied: the food product has to be treated with temperatures close to the boiling point of the food itself. No substantial changes should be observed when speaking of sensorial and nutritional properties, on condition that the processing temperature is close but slightly lower than the maximum thermal value (boiling point of the solid or liquid mixture). For these reasons, the continuous monitoring of rheological properties is required because of the necessity of eliminating possible foams and the consequent local heating (imperfect heat circulation and steam removal). The control of pressure is also recommended: under partial vacuum, boiling points are lowered and the removal of steam is easy at low temperatures. Generally, this process (increase of thermal values) may damage thermally sensible products such as milks, alcoholic beverages and juices.

The second dehydration process, cryoconcentration, is similar to classical freezing systems: liquid foods (solid products cannot be discussed here) are treated with the aim of removing water in form of ice crystals (final concentration of solutes: up to 60 %, while the normal evaporation process should assure only 30 %). The crystallisation of water molecules is carried out by treating the food between -3 and -7 °C: this procedure aims to favour the growth of ice crystals instead of the simple nucleation (Sect. 2.2.1.2). Consequently, food damages (in liquid foods) are limited and the removal of ice crystals is more efficient. Generally, cryoconcentration procedures are used for tea and spices (before drying), coffee extracts and similar products, vinegar, fruit juices, milk and wines. Clearly, the following advantages are determined by the notable removal of volumes (and weights):

– Excellent aroma retention
– Different logistic solutions during transportation and storage steps in the food chain
– Virtual absence of sensory variations and damages to certain organic substances (vitamins).

On the other side, the incorrect management of processing parameters and the insufficient water removal could cause different defects, including microbial spreading and possible phase separations in the liquid product.

The third concentration procedure allows the partial water removal by means of the use of artificial membranes (concentration of solutes may reach 50 %). Interestingly, this removal can easily be performed at room temperature, without classical thermal damages (nutritional loss, variation of sensorial features, etc.). One of the most known versions of concentration by membrane processes, the direct osmosis, is based on the contact (13.5 ± 1.5 h) between whole, cut or sliced foods on the one side and a (possibly heated) concentrated solution (65–70 % of sucrose). In this situation, cellular membranes in foods provide the physical interface between foods and the concentrated solution. The addition of food additives such as ascorbic acid may be recommended because of possible browning effects.

Naturally, this method is completely 'natural'; on the other hand, sweet tastes should be notably enhanced in the final product. A possible countermeasure is the concomitant addition of salt (it can also reduce browning effects and increase osmotic pressures).

Anyway, direct osmosis is not a standalone preservation treatment: on the contrary, it should be recommended as a preliminary process before other drastic preservation procedures (drying, freezing and lyophilisation). In addition, recent modifications of the original system imply the use of relatively short times (up to 20 min) and high temperatures (75 ± 10 °C) with the declared aim of combining synergically dehydration and pasteurisation. At present, the new frontier is represented by the use of artificial membranes (organic polymers): new and improved processes are named 'reverse osmosis' systems.

2.4.2.2 Drying

Drying is historically well known: this method implies the heat treatment of solid or liquid products with the aim of removing large water amounts: the final aqueous content should not exceed 15 %. Differences between historical and modern systems include the following points:

(a) Heat sources. Ancient systems imply the exposure of foods to sun and air for weeks or months. Industrial systems, surely more manageable than historical procedures, are based on the development of artificially heated rooms and the subdivision of the whole process in three steps: food preparation before drying, drying and final packaging
(b) Drying procedures. Historical systems do not require peculiar standards when speaking of monitoring and controls. On the other hand, modern systems—hot air drying; infrared, ultrasound or microwave drying; direct food contact drying—require the critical control of heat transfer from emitting sources to the product and the concomitant water removal until a new thermohygrometric equilibrium with the surrounding environment. Other critical parameters are the gradient heating of foods (core temperatures should not exceed 70 °C), the application of very low pressures and the management of relative humidity (desired values between 10 and 20 %). Insufficient controls could cause serious damages to foods, similarly to failures observed in certain thermal processes (pasteurisation, sterilisation).

2.4.2.3 Lyophilisation

Lyophilisation consists in the under-vacuum dehydration of previously frozen products by sublimation (four steps are required). Temperature has to be <0 °C. Treated foods have a reduced weight (final values can reach 92 %): rehydrated products can show the virtual absence of nutritional losses. In detail, the following products—high-protein and high-calorie soups, granular and orange juice, vegetable soups, meat, etc.—can be preserved for very long times. On the other hand, the main disadvantage of these systems is correlated with the incorrect management of these processes and the hygroscopic behaviour of treated foods: packaging processes should be carried out packaged under vacuum or under nitrogen in triple-layer envelopes.

2.4.2.4 Smoking

Smoking treatments mean the introduction of peculiar substances in foods by means of the simple immersion in a saturated atmosphere or in a liquid bath. Basically, these compounds can:

(a) Penetrate the inner structure of treated foods towards the centre of the product (depending on immersion time) by food contact on edible surfaces with the 'smoking' source
(b) Remain into the food after immersion
(c) Inhibit microbial growth during immersion and after the treatment (prolonged effects).

Smoking procedures are generally performed on already salted foods: the synergic action of salting and smoking can increase dehydration and enhance peculiar aroma and taste features.

From the technological viewpoint, smoking is recognised to show antiseptic, antioxidant and antimicrobial properties. Historically, foods were smoked by hanging over burning wood: as a result, the action of smoking was associated with heat penetration and superficial drying. At present, modern operations have different processing parameters. In detail, two main industrial applications are available:

(1) 'Cold' smoking. foods are heated between 20 and 45 °C; total processing times are measurable as days or weeks
(2) 'Hot' smoking. Intermediate products are heated between 50 and 90 °C for a few hours only.

Because of the nature of smoking compounds, sensorial features of treated foods are completely different from original raw materials or intermediates. As a result, smoked products are extremely appreciated as a new food category, both for hedonistic reasons and historical features. The following list shows the most known types of smoked food products:

(a) Cheeses such as palmero, provola, ewe cheese, etc. (Guillén et al. 2007; Majcher et al. 2011; Naccari et al. 2008)
(b) Cured, cooked and raw meats, such as beef muscle cooked in brine, wurstel, sausages, Prague ham, Speck, etc. (Budig 2012; Lucarini et al. 2013)

(c) Fish (salmon, herring, sturgeon, swordfish, tuna, mackerel, trout and eel). Interestingly, smoking treatments are associated with a certain increase in sodium amounts.

From the technological viewpoint, it should be considered that smoked products are certainly dry versions of the original foods; for this reason, durabilities are notably enhanced (the influence of salting has to be taken into account).

On the other hand, smoking is questioned enough at present because of certain safety concerns (presence of potentially carcinogenic substances such as benzopyrene). As a result, the food industry is currently evaluating smoke flavourings or 'liquid smoke' solutions. In particular, the liquid solution corresponds to the result of the condensation of volatile smoke products (from wood combustion) in water; after a filtration process, the liquid smoke is free from heavy oils and ready for use.

New smoking systems may assure other advantages if compared with traditional methods, including a notable uniformity with relation to flavours, aroma and colours on food surfaces. Moreover, processing times are remarkably faster. Disadvantages of these methods, typically caused by the improper management of times and/or the incorrect dilution of the initial smoke in the final water solution, might be worsened by salting and thermal preservation systems. In particular, excessive salting may enhance crust formation in certain cheeses, and smoking treatments could have more visible effects (brownish colours) than expected; the same thing may occur if preheating treatments give more or less concentrated products, with concern to moisture amounts.

2.4.2.5 Fermentation

Fermentation means the process—or the sum of concomitant processes—that allows certain microorganisms, including degradative life forms such as moulds and yeasts, to break down carbohydrates with the production of alcohol and organic acids. These substances can efficiently inhibit the growth of putrefactive microorganisms and increase shelf life periods. In addition, fermentative processes can modify partially chemical compositions and sensorial features, the digestibility of preserved foods, and other properties. At present, the role of biotechnology is relevant when speaking of industrial fermentation processes: more than 3500 different types of fermented foods are known and produced worldwide, including yoghurts, cheeses, wines, beers and sauerkrauts (Baglio 2014).

Apparently, fermentation can give a number of advantages when speaking of food safety and product categorisation. However, there are some risks because fermentative processes can assure good performances on condition that 'food supports' are good enough. In other words, heat-based thermal treatments have to destroy all degradative microorganisms with the aim of eliminating all possible causes of microbial competition between fermentative life forms and other microbial agents; should not this condition be respected, raw materials could give unexpected results. At the same time, strongly pasteurised raw materials could have a reduced amount of fermentative life forms, and the normal fermentation process could be seriously impeded. With relation to cold-based storage treatments, several microorganisms might suffer important damages and be inhibited in terms of fermentation. Finally, drastic heat-based preservation treatments could decrease the bioavailability of certain nutrients, with comprehensible effects on the normal fermentation yield.

References

Ahmed SH, Luksas AJ (1988) Spreadable butter-like composition and method for production thereof. US Patent 4,769,255, 6 Sept 1988

Alais C (1984) Science du Lait. Principes des techniques laitières, 4th edn. S.E.P.A.I.C., Paris

Ali S, Bettelheim FA (1985) Non-freezing water in protein solutions. Colloid Polym Sci 263 (5):396–398. doi:10.1007/BF01410387

Andersen HJ, Bertelsen G, Ohlen A, Skibsted LH (1990) Modified packaging as protection against photodegradation of the colour of pasteurized, sliced ham. Meat Sci 28(1):77–83. doi:10.1016/0309-1740(90)90021-W

Andreini R, Chiodi J, Noni ID, Resmini P, Battelli G, Cecchi L, Todesco R, Cattaneo TMP, Rampilli M, Foschino R (1990) Evaluation of heat damage in UHT and in-bottle sterilized milk samples traded in Italy. Sci Tec Latt Casearia 41(6):472–492

Baglio E (2014) The modern yoghurt: introduction to fermentative processes. In: Baglio E (ed) Chemistry and technology of Yoghurt Fermentation. SpringerBriefs in Chemistry of Foods, Springer International Publishing, Cham. doi:10.1007/978-3-319-07377-4_1

Borde C (2006) La Revolution sardiniere: Pecheurs et conserveurs en Bretagne sud au XIXe siecle (review). Technol Cult 47(1):200–201. doi:10.1353/tech.2006.0058

Brant AW, Patterson GW, Walters RE (1968) Batch pasteurization of liquid whole egg. 1. Bacteriological and functional property evaluation. Poult Sci 47(3):878–885. doi:10.3382/ps.0470878

Brunazzi G, Parisi S, Pereno A (2014) Packaging and quality. In: Brunazzi G, Parisi S, Pereno A (eds) The importance of packaging design for the chemistry of food products. Springer International Publishing, Cham. doi:10.1007/978-3-319-08452-7_5

Bruhn CM, Schutz HG (1999) Consumer food safety knowledge and practices. J Food Saf 19 (1):73–87. doi:10.1111/j.1745-4565.1999.tb00235.x

Budig J (2012) Prague ham: the past and the present. Maso Int J Food Sci Technol 1:77–80. http://www.maso-international.cz/category/volume-012012/. Accessed 28 Apr 2016

Chandan RC (2015) Dairy processing and quality assurance: an overview. In: Chandan RC, Kilara A, Shah NP (eds) Dairy processing and quality assurance. Wiley, Chichester. doi:10.1002/9781118810279.ch01

Cinquanta L, Albanese D, Cuccurullo G, Di Matteo M (2010) Effect on orange juice of batch pasteurization in an improved pilot-scale microwave oven. J Food Sci 75(1):E46–E50. doi:10.1111/j.1750-3841.2009.01412.x

Clare DA, Bang WS, Cartwright G, Drake MA, Coronel P, Simunovic J (2005) Comparison of sensory, microbiological, and biochemical parameters of microwave versus indirect UHT fluid skim milk during storage. J Dairy Sci 88(12):4172–4182. doi:10.3168/jds.S0022-0302(05)73103-9

Corzo N, López-Fandiño R, Delgado T, Ramos M, Olano A (1994) Changes in furosine and proteins of UHT-treated milks stored at high ambient temperatures. Z Lebensm Unters Forsch 198(4):302–306

Cunningham FE (1986) Egg-product pasteurization. In: Stadelman WJ, Cotteril OJ (eds) Egg science and technology, 3rd edn. P. Macmillan, London, pp 243–272

Datta N, Elliott AJ, Perkins ML, Deeth HC (2002) Ultra-high-temperature (UHT) treatment of milk: comparison of direct and indirect modes of heating. Aust J Dairy Technol 57(3):211–227

Degraeve P, Saurel R, Coutel Y (2003) Vacuum impregnation pretreatment with pectin-methylesterase to improve firmness of pasteurized fruits. J Food Sci 68(2):716–721. doi:10.1111/j.1365-2621.2003.tb05738.x

Delgado A, Vaz de Almeida MD, Parisi S (2016) Chemistry of the Mediterranean diet. Springer International Publishing, Cham. doi:10.1007/978-3-319-29370-7

Delia S, Laganà P, Parisi S (2005) Materiali e metodi di confezionamento nella conservazione dei prodotti alimentari refrigerati. Proceedings of the National Conference "Microbiologia degli alimentai conservati in stato di refrigerazione". Oxoid S.p.A., Garbagnate M.se, pp 93–111

Donati C (2015) Risk analysis, contaminants and impact on health in imports of non-animal origin: the EU context. In: Montanari F, Jezso V, Donati C (eds) Risk regulation in non-animal food imports. SpringerBriefs in Chemistry of Foods, Springer International Publishing, Cham. doi:10.1007/978-3-319-14014-8_1

Engelman MS, Sani RL (1983) Finite-element simulation of an in-package pasteurization process. Numer Heat Transf 6(1):41–54. doi:10.1080/01495728308963073

Ercolini D, Russo F, Torrieri E, Masi P, Villani F (2006a) Changes in the spoilage-related microbiota of beef during refrigerated storage under different packaging conditions. Appl Environ Microbiol 72(7):4663–4671. doi:10.1128/AEM.00468-06

Ercolini D, La Storia A, Villani F, Mauriello G (2006b) Effect of a bacteriocin-activated polyethylene film on *Listeriamonocytogenes* as evaluated by viable staining and epifluorescence microscopy. J Appl Microbiol 100(4):765–772. doi:10.1111/j.1365-2672.2006.02825.x

Farouk MM, Wieliczko KJ, Merts I (2004) Ultra-fast freezing and low storage temperatures are not necessary to maintain the functional properties of manufacturing beef. Meat Sci 66(1):171–179. doi:10.1016/S0309-1740(03)00081-0

Feist EM, Hadi HT (1965) Liquefaction of gases. US Patent 3,224,207, 21 Dic 1965

Fry DE (2001) In vino veritas. Surg Infect 2(3):185–191. doi:10.1089/109629601317202669

Guillén MD, Palencia G, Sopelana P, Ibargoitia ML (2007) Occurrence of polycyclic aromatic hydrocarbons in artisanal Palmero cheese smoked with two types of vegetable matter. J Dairy Sci 90(6):2717–2725. doi:10.3168/jds.2006-452

Samani BH, Khoshtaghaza MH, Minaei S, Zareifourosh H, Eshtiaghi MN, Rostami S (2016) Design, development and evaluation of an automatic fruit-juice pasteurization system using microwave–ultrasonic waves. J Food Sci Technol 53(1):88–103. doi:10.1007/s13197-015-2026-6

Hugas M, Garriga M, Monfort JM (2002) New mild technologies in meat processing: high pressure as a model technology. Meat Sci 62(3):359–371. doi:10.1016/S0309-1740(02)00122-5

Kaale LD, Eikevik TM, Rustad T, Kolsaker K (2011) Superchilling of food: a review. J Food Eng 107(2):141–146. doi:10.1016/j.jfoodeng.2011.06.004

Karabudak E, Bas M, Kiziltan G (2008) Food safety in the home consumption of meat in Turkey. Food Control 19(3):320–327. doi:10.1016/j.foodcont.2007.04.018

Kröckel L (1995) Bacterial fermentation of meats. In: Campbell-Platt G, Cook PE (eds) Fermented meats. Springer US, New York, pp 69–109. doi:10.1007/978-1-4615-2163-1_4

International Dairy Foods Association (2006) Pasteurization. http://www.idfa.org/news-views/media-kits/milk/pasteurization. Accessed 27 Apr 2016

Larousse J (1991) Evolution technologique de l'appertisation et de la pasteurisation des fruits et legumes. Ind Aliment Agric 108(6):475–478

Li B, Sun DW (2002) Novel methods for rapid freezing and thawing of foods–a review. J Food Eng 54(3):175–182. doi:10.1016/S0260-8774(01)00209-6

Lucarini M, Saccani G, D'Evoli L, Tufi S, Aguzzi A, Gabrielli P, Marletta L, Lombardi-Boccia G (2013) Micronutrients in Italian ham: a survey of traditional products. Food Chem 140(4):837–842. doi:10.1016/j.foodchem.2012.10.020

Magnussen OM, Haugland A, Hemmingsen AKT, Johansen S, Nordtvedt TS (2008) Advances in superchilling of food–Process characteristics and product quality. Trends Food Sci Technol 19 (8):418–424. doi:10.1016/j.tifs.2008.04.005

Mangalassary S, Dawson PL, Rieck J, Han IY (2004) Thickness and compositional effects on surface heating rate of bologna during in-package pasteurization. Poult Sci 83(8):1456–1461. doi:10.1093/ps/83.8.1456

Mangalassary S, Han I, Rieck J, Acton J, Dawson P (2008) Effect of combining nisin and/or lysozyme with in-package pasteurization for control of Listeria monocytogenes in ready-to-eat turkey bologna during refrigerated storage. Food Microbiol 25(7):866–870. doi:10.1016/j.fm. 2008.05.00

Majcher MA, Goderska K, Pikul J, Jeleń HH (2011) Changes in volatile, sensory and microbial profiles during preparation of smoked ewe cheese. J Sci Food Agric 91(8):1416–1423. doi:10. 1002/jsfa.4326

Mauriello G, Ercolini D, La Storia A, Casaburi A, Villani F (2004) Development of polyethylene films for food packaging activated with an antilisterial bacteriocin from *Lactobacillus curvatus* 32Y. J Appl Microbiol 97(2):314–322. doi:10.1111/j.1365-2672.2004.02299.x

Micali M, Parisi S, Minutoli E, Delia S, Laganà P (2009) Alimenti confezionati e atmosfera modificata. Caratteristiche basilari, nuove procedure, applicazioni pratiche. Ind Aliment 48 (489):35–43

Molin G (2000) Modified atmospheres. In: Lund BM, Baird-Parker TC, Gould GW (eds) The microbiological safety and quality of food, vol 1. Aspen Publishers Inc, Gaithersburg

Morales FJ, Romero C, Jiménez-Pérez S (2000) Characterization of industrial processed milk by analysis of heat-induced changes. Int J Food Sci Technol 35(2):193–200. doi:10.1046/j.1365-2621.2000.00334.x

Muftugil N (1986) Effect of different types of blanching on the color and the ascorbic acid and chlorophyll contents of green beans. J Food Proc Preserv 10(1):69–76. doi:10.1111/j.1745-4549.1986.tb00006.x

Murphy RY, Duncan LK, Driscoll KH, Marcy JA (2003) Lethality of Salmonella and Listeria innocua in fully cooked chicken breast meat products during postcook in-package pasteurization. J Food Prot 66(2):242–248

Murphy RY, Hanson RE, Johnson NR, Chappa K, Berrang ME (2006) Combining organic acid treatment with steam pasteurization to eliminate Listeria monocytogenes on fully cooked frankfurters. J Food Prot 69(1):47–52

Murray M, Richard JA (1997) Comparative study of the antilisterial activity of nisin A and pediocin AcH in fresh ground pork stored aerobically at 5°C. J Food Prot 60:1534–1540

Naccari C, Cristani M, Licata P, Giofrè F, Trombetta D (2008) Levels of benzo [a] pyrene and benzo [a] anthracene in smoked "Provola" cheese from Calabria (Italy). Food Addit Contam 1 (1):78–84. doi:10.1080/19393210802236968

Negi PS, Roy SK (2000) Effect of blanching and drying methods on β-carotene, ascorbic acid and chlorophyll retention of leafy vegetables. LWT-Food Sci Technol 33(4):295–298. doi:10.1006/fstl.2000.0659

Ouwehand AC (1998) Antimicrobial components from lactic acid bacteria. In: Salminen S, von Wright A (eds) Lactic acid bacteria: microbiology and functional aspects, 2nd edn. Marcel Dekker Inc., New York, pp 139–159

Parisi S (2002) I fondamenti del calcolo della data di scadenza degli alimenti: principi ed applicazioni. Ind Aliment 41(417):905–919

Parisi S (2005) La produzione "continua" è anche "costante"? Confutazione di alcuni luoghi comuni nel settore industriale/manifatturiero. Chim Ital 16(3–4):10–18

Parisi S (2009) Intelligent packaging for the food industry. In: Carter EJ (ed) Polymer electronics —a flexible technology. Smithers Rapra Technology Ltd, Shawbury

Parisi S (2012) Food packaging and food alterations. Smithers Rapra Technology Ltd, Shawbury

Parisi S (2013) Food Industry and packaging materials: performance-oriented guidelines for users. Smithers Rapra Technology Ltd, Shawbury

Parisi S (2016) Alberto Escarpa, María Cristina González, and Miguel Angel López: Agricultural and food electroanalysis. Anal Bioanal Chem 408(9):2185–2186. doi:10.1007/s00216-016-9347-9

Parisi S, Laganà P, Delia S (2004) Cubed Mozzarella Cheese in modified atmosphere packages: evolutive profiles of chemical and microbiological parameters during Shelf Life. In: Programme & Abstracts Book of the 3rd ILSI Europe International Symposium on Food Packaging, Barcelona, 17–19 Nov 2004

Patterson MF (2005) Microbiology of pressure-treated foods. J Appl Microbiol 98(6):1400–1409. doi:10.1111/j.1365-2672.2005.02564.x

Potter NN, Hotchkiss JH (1995) Milk and milk products. Food Science. Springer, US, New York, pp 279–315

Pruthi JS (1999) Quick freezing preservation of foods: foods of plant origin, vol 2. Allied Publishers Ltd, New Delhi

Rahman MS (ed) (2007) Handbook of food preservation, 2nd edn. CRC Press LLC, Boca Raton

Redmond EC, Griffith CJ (2003) Consumer food handling in the home: a review of food safety studies. J Food Prot 66(1):130–161

Reif-Acherman S (2009) Several motivations, improved procedures, and different contexts: the first liquefactions of helium around the world. Int J Refrig 32(5):738–762. doi:10.1016/j.ijrefrig.2009.02.019

Rerkrai S, Jeon IJ, Bassette R (1987) Effect of various direct ultra-high temperature heat treatments on flavor of commercially prepared milks. J Dairy Sci 70(10):2046–2054. doi:10.3168/jds.S0022-0302(87)80252-7

Rigo M (2010) Sistemi di pastorizzazione a radiofrequenza. Dissertation, University of Padua

Rogers WR (2013) Background to sterilisation—an historical introduction. In: Rogers WR (ed) Healthcare sterilisation: introduction and standard practices, vol 1. Smithers Rapra Technology, Shawbury

Schubring GR (2009) "Superchilling"—an "old" variant to prolong shelf life of fresh fish and meat requicked. Fleischwirtsch 89(9):S.104–113

Senesi E (1984) Surgelazione dei prodotti vegetali. In: Ottaviani F (ed) Microbiologia dei prodotti di origine vegetale – ecologia ed analisi microbiologica. Chiriotti Editori, Pinerolo

Senesi E, Bignardi B, Gennari M (2000) Influence of pre-treatment and protective atmosphere packaging on quality indices of fresh-cut musk melon. Fruttic 11:69–73

Simpson BK, Rui X, Klomklao S (2012) Enzymes in food processing. In: Simpson BK (ed) Food biochemistry and food processing, Second Edition. Wiley-Blackwell, Oxford. doi:10.1002/9781118308035.ch9

Sivertsvik M, Jeksrud WK, Rosnes JT (2002) A review of modified atmosphere packaging of fish and fishery products–significance of microbial growth, activities and safety. Int J Food Sci Technol 37(2):107–127

Skandamis PN, Nychas GJE (2001) Effect of oregano essential oil on microbiological and physico-chemical attributes of minced meat stored in air and modified atmospheres. J Appl Microbiol 91(6):1011–1022. doi: 10.1046/j.1365-2672.2001.01467.x

Soncini G (2008) Frodi e alterazioni delle conserve. Proceedings of the Fourth 'Meeting Nazionale sulla Sicurezza Alimentare: le Frodi Alimentari e le Alterazioni degli Alimenti', Rome, 23–24 May 2008

Stabel JR (2001) On-farm batch pasteurization destroys Mycobacterium paratuberculosis in waste milk. J Dairy Sci 84(2):524–527. doi:10.3168/jds.S0022-0302(01)74503-1

Tassou C, Koutsoumanis K, Nychas GJE (2000) Inhibition of Salmonella enteritidis and Staphylococcus aureus in nutrient broth by mint essential oil. Food Res Int 33(3–4):273–280. doi:10.1016/S0963-9969(00)00047-8

Waterman JJ, Taylor DH (2001) Superchilling. Torry Advisory Note No. 32. Torry Research Station, UK. FAO in partnership with Support Unit for International Fisheries and Aquatic Research, SIFAR

Wheeler JL, Harrison MA, Koehler PE (1987) Presence and stability of patulin in pasteurized apple cider. J Food Sci 52(2):479–480. doi:10.1111/j.1365-2621.1987.tb06643.x

Zhao Y (2007) Freezing process of berries. In: Zhao Y(ed) Berry fruit: value-added products for health promotion. CRC Press, Boca Raton

Chapter 3
Undesired Chemical Alterations and Process-Related Causes. The Role of Thermal Control and the Management of Thermal Machines

Marco Fiorino and Salvatore Parisi

Abstract This chapter is explicitly dedicated to the description of several known sensorial alterations in foods and beverages because of technological preservation systems. Many possible alterations could be considered: the more diversified the choice of industrial processes (including also traditional preservation systems), the broader the field of interest. For this reason, authors have decided to consider the subcategory of sensorial failures in relation to thermally conducted processes only. Basically, the following chemical alterations are discussed in relation to thermal food processing: hydrolytic decomposition of fat molecules and proteins; ketonic and oxidative rancidity; homofermentative reactions; decomposition and modification of lateral chains in amino acids; protein denaturation; interactions between proteins and other organic molecules; enzymatic degradation of proteins; caramelisation of carbohydrates.

Keywords Denaturation · Enzymatic degradation · Food preservation · Hydrolysis · Mechanical process · Rancidity · Thermal dissipation

Abbreviations

UHT Ultra High Temperature
a_w Water activity

M. Fiorino (✉)
Technical Consultant, Siracusa, Italy
e-mail: marco.fiorino.ingegnere@gmail.com

S. Parisi
Industrial Consultant, Palermo, Italy
e-mail: drparisi@inwind.it

© The Author(s) 2016
M. Micali et al., *The Chemistry of Thermal Food Processing Procedures*,
Chemistry of Foods, DOI 10.1007/978-3-319-42463-7_3

3.1   Introduction to Chemical and Physical Modifications of Foods in the Modern World

One of the most interesting (and worrying) features of the modern food and beverage industry is correlated with the irreversible modification of certain chemical–physical parameters of produced foods. Actually, the variation of sensorial features in edible products is not a recent discovery (Cappelli and Vannucchi 1990); the same thing can be affirmed when speaking of microbiological alterations caused by the modification of edible 'culture media' (foods, beverages) after processing techniques (Parisi 2002, 2003). However, the current situation appears different if compared with the past: in other words, the increasing industrialisation worldwide has certainly modified the organoleptic representation of current foods and beverages (Parisi 2006, 2012, 2013) with partially forecasted results on the one hand and the concomitant confusion of selected brands and related features by normal consumers on the other side.

Anyway, many different chemical and physical modifications may be observed when speaking of industrial and artisanal foods and beverages. A preliminary discrimination should concern the frequency of certain variations on the chemical level, because an industrialised process is expected to modify intermediate edible masses from the start to the end of the whole 'chain' in a predictable way. In other words, the more standardised the process, the lower the amount of unexpected 'surprises' in the final product. On the other side, artisanal products—the heritage of old times and civilisations—are easily recognised as the carrier of many unexpected or unknown sensorial variations when discriminating between two or more similar products of different geographical or social origin (Gómez-Ruiz et al. 2002; Kupiec and Revell 1998; Tregear 2003).

By a general viewpoint, the industrial food product (Campbell 2009) is designed, produced and distributed with the aim of 'capturing' the attention of 'normal consumers' (Parisi 2004–2012) in a durable manner. For this reason, each product has to be obtained with the minor possible amount of sensorial variations when speaking of different production lots: different results could easily disorientate consumers. In fact, the simple and inevitable modification of sensorial features in food products is irreversible between the start and the end of the commercial life (within the so-called expiration or sell-by-date). Consequently, food technologists should compose their personalised idea of food or beverage product taking inevitable temporal variations of colours, aroma, taste, texture and shape into account. The occurrence of additional modifications has to be avoided. On the other side, artisanal foods are surely restricted in the modern world when speaking or market sales: the competition may be assured if a sort of shared organisation is put in place. Normally, these joint entities may assure elements such as legal protection of brands, communications, innovation and quality assurance (Jordana 2000). The last feature, quality assurance, should imply the good or acceptable knowledge of technological processes and related effects, including undesired alterations (or 'surprises').

Food processing and related alterations

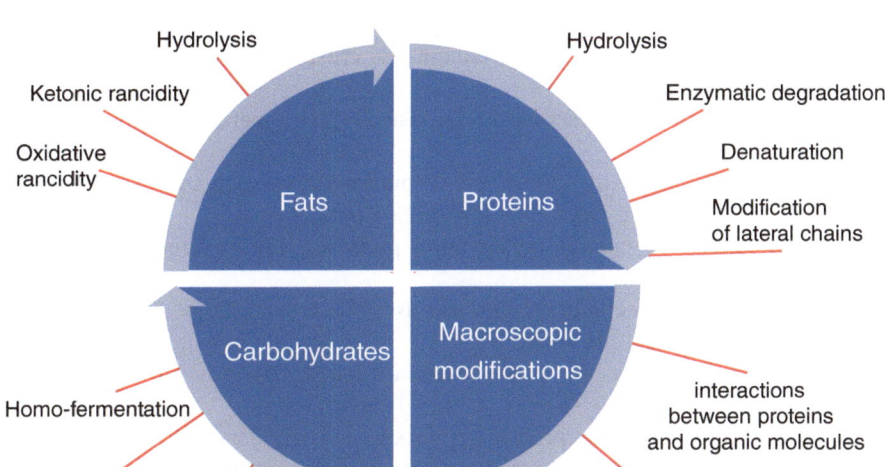

Fig. 3.1 A brief overview of chemical alterations during food processes. In spite of the possible influence of other causes, main chemical and physical alterations of foods and beverages are often originated during processing steps in the food industry, and during the remaining part of the whole food chain (storage, delivery and final distribution). Generally, the following procedures and reactions are considered when speaking of probable causes: controlled enzymatic reactions; controlled seasoning in open air; controlled homo- and hetero-fermentative processes; drying systems; mechanical processes; irradiation; high-temperature preservation; low-temperatures processes; MAP and CAS systems; semi-preservation (addition of salt, ethanol, sucrose, fatty oils, etc.); smoking treatments

This chapter is explicitly dedicated to the description of several known sensorial alterations in foods and beverages because of technological processes. Actually, the discussion is broad because many possible alterations could be considered: the more diversified the choice of industrial processes (including also traditional preservation systems), the broader the field of interest. For this reason, authors have decided to consider the subcategory of organoleptic failures or 'surprises' in relation to thermally conducted processes only (Fig. 3.1). In other words, the discussion does not concern chemical, physical, microbiological, sensorial alterations and synergistic results caused by processes without a clearly correlated thermal augment.

In fact, the following processes or systems may be taken into account when speaking of undesired food or beverage alterations (Cappelli and Vannucchi 1990; Delia et al. 2005; Fessmann 1999; Horner 1997; Micali et al. 2009; Parisi 2009; Rahman and Perera 1999; Šimko 2002):

- Controlled enzymatic reactions
- Controlled seasoning in open air (presence of selected microorganisms)
- Controlled seasoning in open air

- Controlled homo- and hetero-fermentative processes
- Drying systems including cryostatic processes, membrane- and microwave-assisted technologies and freeze-drying
- Mechanical processes (kneading, lamination, crushing, screwing, etc.)
- Radiation techniques
- Thermal systems with preservation purposes, high temperatures: pasteurisation, sterilisation, etc.
- Thermal processes with preservation purposes, use of low temperatures: refrigeration, freezing, etc.
- Thermal processes without preservation purposes, use of high temperatures: cooking, melting, etc.
- Other preservation techniques with the use of active systems and/or modified gases as food-contact atmosphere
- Chemical systems for the preservation of foods: addition of salt, ethanol, sucrose, fatty oils, etc.
- Smoking treatments for preserving foods and similar techniques without preservation meaning, also named 'aromatisation' procedures.

The above-mentioned systems are the basis of the modern industrial production when speaking of foods or beverages. Each processing step or sub-step has its own reason or reasons (advantages) in the ambit of the whole food process. On the other side, the occurrence of several defects may be ascribed to the same group of systems.

In general, the following chemical alterations should be discussed in relation to food processing and other causes:

(a) Hydrolytic decomposition of fat molecules, catalysed by microbial lipases, light exposure and water (Holliday et al. 1997; Macrae 1983; Sherwin 1978)
(b) Ketonic rancidity. Enzymatic reactions are involved; mould contamination is needed (Hamilton 2003; Kinderlerer 1993; Kinderlerer and Kellard 1984)
(c) Oxidative rancidity, catalysed by light exposure, presence of metals such as iron, copper, nickel and other metallic elements, thermal dissipation during mechanical processes, presence of peroxides and lipoxydases (Aluyor and Ori-Jesu 2008; Kochhar 1996; Mizutani and Hashimoto 2004; Tan and Man 2002)
(d) Homofermentative reactions (Marshall and Tamime 1997) with the production of ethanol, and hetero-fermentative processes (Pilone et al. 1991; Priest 1999) with these final products: ethanol, isopropyl alcohol, butyl alcohol, acetone, etc. These reactions are naturally catalysed by different microorganisms such as *Lactobacillaceae*, yeasts and moulds
(e) Decomposition and modification of lateral chains in amino acids (Stadtman and Levine 2003)
(f) Protein denaturation (Hendrickx et al. 1998; Lopez-Fandiño et al. 1996; Messens et al. 1997)
(g) Different interactions between proteins and other organic molecules. 'Maillard reaction' is one of the most known and studied examples (Fu et al. 1994; Martins et al. 2000)

(h) Hydrolysis of proteins (Sarmadi and Ismail 2010)
(i) Enzymatic degradation of proteins (Cappelli and Vannucchi 1990)
(j) Caramelisation of carbohydrates (Claude and Ubbink 2006; Friedman 1996; Izydorczyk 2005)
(k) Chemical–physical modifications such as fat smearing and breaking of emulsions (McCain 1991).

From the sensorial viewpoint, many of the above-mentioned reactions determine important variations. Generally, colours are modified in an irreversible way: the caramelisation and other reactions with carbohydrates as substrates are the cause of brown colours in foods (Cappelli and Vannucchi 1990). Moreover, the following chemicals are modified and/or destroyed:

– Vitamins. In particular, vitamin C, β-carotene and thiamine are progressively destroyed (Morris et al. 2004) because of oxidation, contact with water, and thermal processes such as blanching (vegetable products)
– Anthocyanins (Jackman et al. 1987; Patras et al. 2010)
– Carotenoids (Britton and Khachik 2009; Rodriguez-Amaya 1997).

The physical looking of foods and beverages can be notably modified: as an example, textural variations in several products such as certain cheeses and industrial pasta have been considered in the scientific literature and studied with the aim of providing better products (Parisi et al. 2006). Anyway, the reduction of hydric contents is one of the most known effects when speaking of textural variations and correlated effects (Barbieri et al. 2014). Certain superficial modifications, including also the simple aspect of products (example: geometrical shape) can be also observed and mathematically correlated with chemical–physical parameters. Finally, the aroma may be remarkably modified with pleasant or unpleasant results, depending on the process; naturally, alterative phenomena (degradation) can also determine the premature perishability of certain foods or beverages with claimed long durability. Should this situation be observed, production/storage/delivery steps and the distribution process (or the sum of independent and subsequent sub-steps in the food chain) would be studied carefully.

Anyway, the sum of possible modifications is very large and this book could not consider all options here. For this reason, chemical–physical modifications are discussed here if some food processing relationship is involved and the process tends to the increase of thermal values. In addition, some reflection is made here with concern to the importance of thermal control and the management of involved machinery and equipments in the modern food industry.

In general, the following list can be exhaustive enough when speaking of thermal augment:

• Mechanical processes with heat dispersion (lamination, extrusion, kneading, grinding, crushing, screwing, mixing, etc.)
• Thermal systems with preservation purposes, use of high temperatures: pasteurisation, sterilisation, etc.

- Thermal processes without preservation purposes, use of high temperatures: cooking, melting, etc.

It should be noted that mechanical processes are not carried out with the aim of increasing thermal values of intermediate food masses. On the other hand, the simple mechanical movement determines a certain energy dissipation because of compression, impact and friction forces, with the consequent heat dispersion into working masses (Hernandez-Izquierdo and Krochta 2008; Zardetto and Dalla Rosa 2006). This heat dissipation is usually observed in grinding operations on dry foods (coffee, cocoa, etc.), although other sectors such as meat processing industries may show similar problems (McCain 1991). For this reason, mechanical processes should be considered in this ambit despite the low impact in terms of thermal increase. On the other side, it has to be considered that thermal increases may become notable during the process with the consequent melting of grinded particles. Moreover, heat exposure of initial raw materials (rice) during grinding has been reported to have apparently some influence on the quality of finished products (Nishita and Bean 1982). Cryogens such as carbon dioxide snow are often used with the aim of reducing thermal augments.

3.2 Undesired Chemical Alterations and Thermal Processing Management

In general, each food processing technique has one, two or three distinct aims:

(a) The safe consumption of foods or beverages
(b) The easy preservation of claimed long-durability foods and beverages
(c) The increase of palatability, for peculiar foods (Fujimoto et al. 2015; Lepper 2013).

Basically, all thermal treatments including the simple cooking process can determine the following situations (Cappelli and Vannucchi 1990):

(1) Reduction of hydric amounts (in other words, augment of dry content in treated foods)
(2) Hydrolysis of proteins, carbohydrates and fatty molecules
(3) Inactivation of many enzymes (microbial origin) and toxins, elimination of a notable part of pathogen microorganisms and reduction of the total viable count of microbial agents in treated foods
(4) Demolition of certain amino acids, thermally unstable vitamins and organic compounds (proteins and carbohydrates) because of different reaction chains (Maillard reaction, etc.)
(5) Aqueous dissolution of mineral salts in the excess water (after treatment) with consequent diminution of nutritional values.

Therefore:

(a) Each thermally treated food loses a variable amount of the original weight, depending on its composition, treatment conditions (temperature and time cycles), and the dimension of treated foods and 'cooking' chambers

(b) Certain colorimetric variations can be easily observed because the original aqueous solid solution (the initial food) has been physically concentrated after treatment. As an example, the variation of colours can be easily ascribed to certain fatty molecules without the diluting effect of dissolving water

(c) The digestibility of main nutritional molecules—carbohydrates, proteins and fats—is notably increased, in spite of the presence of certain indigestible fractions

(d) Sensorial features are remarkably enhanced after treatment

(e) The nutritional value may be increased after treatment when speaking of available proteins, fats and carbohydrates; on the other side, vitamins and other nutritive principles are lost

(f) The presence of mineral salts may be reduced after treatment because of the dissolution in the existing (dissolving) water and/or in added cooking waters. This modification may be observed with the concomitant variation of chemical–physical properties of certain macromolecules such as proteins.

The category of thermal processes involves many techniques, and cooking is only the first and best known of these methods. With relation to the industry of vegetable foods, the 'blanching' process (cooking in boiling water, followed by the sudden 'shocking' step under cold water) can be extremely useful (Abu-Ghannam and Jaiswal 2015; Menon et al. 2015). A notable part of thermal processes are carried out with the aim of destroying microorganisms and inactivating toxins with a reliable degree of food safety: pasteurisation, sterilisation and related variations such as ultra high temperature (UHT) processing are widely used in the modern food industry.

Because of the importance of involved parameters on the management of thermal processes and related machinery, a brief description of the above-mentioned processes has been shown in Chap. 2.

However, the main objective of this chapter is the definition of undesired chemical and physical modifications in foods in strict relation to temperature, the main process parameter. For this reason, the above-mentioned modifications are discussed in Sect. 3.3 with reference to thermal values, where available. Before continuing, a little premise should be made with concern to mechanical treatments (kneading, lamination, crushing, screwing, etc.) and their possible effects on undesired chemical reactions.

3.3 Thermal Processing Parameters and Chemistry of Foods

3.3.1 Reduction of Hydric Amounts and Possible Damages After Cooking

Generally, the reduction of moisture in edible products is desired because the microbial spoilage is direct function of liquid water amounts. Actually, the bioavailable water corresponds to a (notable) fraction of the whole water amount that could be analytically determined. This fraction, named 'free water', is generally measured in an indirect way by means of another parameter: water activity (a_w). Moreover, many properties of foods—rheological behaviour, hygroscopicity, etc.—depend strictly on the physical state of water, and the aqueous solvent is not always liquid.

Normally, a_w for a selected food corresponds to the relative humidity of the air above the food surface, on condition that the food is really in equilibrium with the atmosphere. The basic relation is:

$$a_w = \frac{P_0}{P} \qquad (3.1)$$

where P_0 is the vapour pressure for the food (a solution!) and P corresponds to the vapour pressure of pure water at the temperature of measurement (Walstra 2003). The importance of a_w values is correlated with the necessity of eliminating or reducing the microbiological risk, in the ambit of food safety measures (Delia et al. 2005; Ottaviani 2002; Parisi 2002). Anyway, the limit for a safe preservation of foods without the addition of chemical additives (as dissolved substances) or thermal treatments such as pasteurisation corresponds to $a_w \leq 0.65$ (Cappelli and Vannucchi 1990). Below this value, only saccharophilic yeasts could colonise foods.

Apart this consideration, it has been reported that normal reactions in foods have a different speed (in terms of kinetics) depending on the water content. In other words, the lower the amount of available water, the lower the speed of many possible reactions is; including biochemical processes (Walstra 2003). The basis of this reflection cannot be simple because many different subprocesses of different (chemical, physical, biological) origin can occur at the same time with different 'weights'.

Anyway, the reduction of moisture (this quantity is generally different from 'water') in foods is normally desired. With exclusive reference to thermal processes with strong augments of temperatures, it can be affirmed that the most used method concerns evaporation (Cappelli and Vannucchi 1990). In detail, the higher the removal of water molecules from a food product, the higher the quantity of produced (and consumed) heat for this work (Walstra 2003). The continuous removal of aqueous content becomes more and more difficult if a_w gets lower (and sorption enthalpy increases). This phenomenon depends on the direct relationship between the remaining amount of water and the low a_w value, and the necessity of supplying continually sorption heat (also intended as sorption enthalpy) and the additional enthalpy of evaporation[1] during the process. For these reasons, drying processes are easier (and get good yields) at increasing temperatures (Cappelli and Vannucchi 1990; Walstra 2003).

The direct reduction of water can be performed by means of direct or indirect systems. Briefly, it can be affirmed that direct techniques concern the use of electric resistances and low pressures (under-vacuum processes), while indirect systems use heat exchangers and additional machinery (placed into reactors) for increasing heat diffusion in the mixture or intermediate mass. The problem of these processed concerns thermally sensible fluids (orange juices, milk, etc.). For this reason, many processes can be subdivided in different steps where the working pressure is constantly reduced with the aim of reducing evaporation temperatures until 40 or 50 °C. However, thermally sensible products cannot be treated in gradient processes: the reduction of evaporation temperatures has to be reached immediately (one step only). The diffusion of heat can be observed as the result of four different mechanisms: conduction, convection, radiation and distillation (Walstra 2003).

Apart chemical damages, the reduction or moisture can affect the textural appearance of products, but this property depends also on the composition of dry residues: proteins, fats and carbohydrates. Other physical modifications concern the rheology of intermediate foods and beverages, and the possible hygroscopicity. With concern to hygroscopic powders, the observed phenomenon corresponds to the normal tendency of the 'food/external atmosphere' system to a new equilibrium, because of the unbalanced ratio between a_w values (air has a higher a_w than foods). Consequently, the problem should be managed in terms of good (or bad) storage in dry warehouses.

Finally, some reflection could be made with relation to several defects (microbial spoilage is not discussed here) without direct connection to water losses. The removal of water produces a more rigid mass with peculiar properties. As a single example, the following defects have been reported with relation to poultry foods (Sams 2001):

(a) Abnormal emersion of fatty substances in certain meat products because of the excessive amount of fats and collagen in the raw materials or because of excessive heating temperatures (normal end-point values: 68.3–73.9 °C). Fat globules break the complex fat/protein matrix
(b) Possible discoloration in emulsified products because the minimum temperature (end-point: 68 °C) has not been reached.

---

[1]This enthalpy is really higher than sorption heath during the process.

3.3.2 Hydrolysis of Proteins, Carbohydrates and Fatty Molecules

Proteins can be easily hydrolysed under high-temperature processes such as cooking; naturally, pasteurisation (temperatures under 100 °C) and sterilisation (example: more than 3 min at 121 °C) procedures may enhance the hydrolysis. In general, this reaction is a good result because of the increase of digestible proteins; anyway, the amount of hydrolised proteins depends strongly from the acidity of food masses.

Fat molecules can also be hydrolised—the industrial procedure is applied at 250 °C (pressure: 50 bar)—but the initial decomposition in glycerol and fatty acids is unfavourable: the obtained glycerol is subsequently decomposed again with the production of undesired acrolein and two water molecules (Cappelli and Vannucchi 1990). Moreover, several fatty acids may be detrimental when speaking of food palatability.

With relation to carbohydrates, starch molecules can be partially hydrolised in absence of water (temperatures should be around 160 °C) with the production of dextrins; following reactions may also produce single maltose molecules, with a notable augment of digestibility. For this reason, the hydrolysis of carbohydrates in foods does not seem important as risk factor.

3.3.3 Demolition of Nutrients. Maillard Reactions and Other Mechanisms

The action of high temperatures may destroy or modify certain amino acids with important effects. One of the most important and studied mechanisms concerns the transformation of cysteine and cystine residues at temperature >100 °C with the production of hydrogen sulphide. Actually, this simple reaction is part of the sum of reactions concerning the vast ambit of denaturation phenomena. Because of the rupture of disulphide bonds in certain proteins, the production of the typical 'rotten eggs' smell highlights the presence of amino acids rich in sulphur (Parisi 2012).

Other situations concern

– The modification of nitrogen-containing amino acids such as asparagines and glutamine with production of ammonia; modified amino acids may also form covalent bonds (Cappelli and Vannucchi 1990)
– Serin modification with the loss of one water molecule
– The oxidation of methionine to methionine sulphoxide (Narayan 1997)
– The production of certain mutagenic heterocyclic amines may be produced under cooking (temperatures >200 °C) when the presence of tryptophan is observed (Sugimura 1985).

The Maillard reaction can be also observed: this reaction starts as the initial condensation of glucose and a protein with the production of Schiff bases; after this step, a number of compounds can be irreversibly obtained with different (parallel and consecutive) reactions. These products include furfural, different aldimines and ketimes, and the group of brown-coloured melanoidins. The Maillard reaction is extremely complicated and the discussion could not be easily carried out in this Chapter. In addition, the influence of temperature values may not be simply expressed with the Arrhenius equation, but with the basic Eyring equation (Martins et al. 2000; Walstra 2003) containing activation enthalpy, entropy and Gibbs energy terms. In addition, it has been reported that the reaction between proteins and reducing carbohydrates may be observed at room and very low temperatures (Narayan 1997). For this reason, the complex of Maillard reactions is not further discussed here.

Moreover, proteins can easily interact during high-temperature processes with the formation of solid cross-linked matrices (Sams 2001): one of the most known examples is the gelation of myofibrillar proteins in processed poultry products.

Finally, it should be remembered that

(a) Mineral salts are easily dissolved into water; because of the removal of the aqueous solvent, the diminution of certain metals (calcium, etc.) can compromise the stability of partially bonded macromolecules such as proteins, with the consequent coagulation and precipitation

(b) The amount of vitamins is notably reduced under drastic conditions; for this reaction, oxidation can easily attack fatty molecules after treatment. In addition, the reaction between certain food proteins and oxidised fatty acids has been recently reported (Narayan 1997). The problem is not the expected oxidation of fatty acids after the destruction of natural antioxidants (vitamins, etc.), but the production of strongly bound complexes between these fatty acids and proteins. The sum of radicalic, cross-linking, oxidation and transfer reactions has been observed to produce similar complex at 90, 60 and 30 °C too (Narayan 1997). Apparently, the role of protein radicals is important; consequently, the reduction of protein damages in high-temperature (100 °C and above) processes seems crucial.

References

Abu-Ghannam N, Jaiswal A (2015) Blanching as a treatment process: effect on polyphenols and antioxidant capacity of cabbage. In: Preedy V (ed) Processing and impact on active components in food. Elsevier/Academic Press, London, pp 35–43

Aluyor EO, Ori-Jesu M (2008) The use of antioxidants in vegetable oils–a review. Afr J Biotechnol 7(25):4836–4842

Barbieri G, Barone C, Bhagat A, Caruso G, Conley ZR, Parisi S (2014) The problem of aqueous absorption in processed cheeses: a simulated approach. In: Barbieri G, Barone C, Bhagat A, Caruso G, Conley ZR, Parisi S (eds) The influence of chemistry on new foods and traditional products, pp 1–17. SpringerBriefs in Chemistry of Foods, Springer International Publishing, Cham. doi:10.1007/978-3-319-11358-6_1

Britton G, Khachik F (2009) Carotenoids in food. In: Britton G, Pfander H, Liaaen-Jensen S (eds) Carotenoids, vol. 5: Nutrition and Health, pp 45–66. Birkhäuser Basel. doi:10.1007/978-3-7643-7501-0_3

Campbell B (2009) Fast food/slow food: the cultural economy of the global food system. In: Wilk (ed) Fast food/slow food—the cultural economy of the global food system. J Roy Anthropol Inst 15, 2: 411–412. doi:10.1111/j.1467-9655.2009.01566_3.x

Cappelli P, Vannucchi V (1990) Chimica degli alimenti. Conservazione e trasformazione. Zanichelli, Bologna

Claude J, Ubbink J (2006) Thermal degradation of carbohydrate polymers in amorphous states: a physical study including colorimetry. Food Chem 96(3):402–410. doi:10.1016/j.foodchem.2005.06.003

Delia S, Laganà P, Parisi S (2005) Materiali e metodi di confezionamento nella conservazione dei prodotti alimentari refrigerati. Proceedings of the XIV Conferenza Nazionale "Microbiologia degli alimenti conservati in stato di refrigerazione", Facoltà di Chimica Industriale, University of Bologna, pp 93–111

Fessmann KD (1999) Process and apparatus for smoking foodstuffs. US Patent 5,910,330, 8 Jun 1999

Friedman M (1996) Food browning and its prevention: an overview. J Agric Food Chem 44 (3):631–653. doi:10.1021/jf950394r

Fu MX, Wells-Knecht KJ, Blackledge JA, Lyons TJ, Thorpe SR, Baynes JW (1994) Glycation, glycoxidation, and cross-linking of collagen by glucose: kinetics, mechanisms, and inhibition of late stages of the Maillard reaction. Diabetes 43(5):676–683. doi:10.2337/diab.43.5.676

Fujimoto Y, Chiba H, Toko ROK (2015) Effects of different heating methods on positional differences in taste, texture, color, and palatability of simmered food: comparison of sensor responses and sensory evaluation results. Sens Mater 27(5):365–375

Gómez-Ruiz JÁ, Ballesteros C, Viñas MÁG, Cabezas L, Martínez-Castro I (2002) Relationships between volatile compounds and odour in Manchego cheese: comparison between artisanal and industrial cheeses at different ripening times. Lait 82(5):613–628. doi:10.1051/lait:2002037

Hamilton RJ (2003) Oxidative rancidity as a source of off-flavours. In: Baigrie B (ed) Taints and off-flavours in food. Woodhead Publishing Limited, Cambridge, pp 140–161

Hendrickx M, Ludikhuyze L, Van den Broeck I, Weemaes C (1998) Effects of high pressure on enzymes related to food quality. Trends Food Sci Technol 9(5):197–203. doi:10.1016/S0924-2244(98)00039-9

Hernandez-Izquierdo VM, Krochta JM (2008) Thermoplastic processing of proteins for film formation—a review. J Food Sci 73(2):R30–R39. doi:10.1111/j.1750-3841.2007.00636.x

Holliday RL, King JW, List GR (1997) Hydrolysis of vegetable oils in sub-and supercritical water. Ind Eng Chem Res 36(3):932–935. doi:10.1021/ie960668f

Horner WFA (1997) Preservation of fish by curing (drying, salting and smoking). In: Hall GM (ed) Fish processing technology. Springer US, New York, pp 32–73. doi:10.1007/978-1-4613-1113-3_2

Izydorczyk M (2005) Understanding the chemistry of food carbohydrates. In: Cui SW (ed) Food carbohydrates—chemistry, physical properties, and applications. CRC Press, Boca Raton

Jackman RL, Yada RY, Tung MA, Speers R (1987) Anthocyanins as food colorants—a review. J Food Biochem 11(3):201–247. doi:10.1111/j.1745-4514.1987.tb00123.x

Jordana J (2000) Traditional foods: challenges facing the European food industry. Food Res Int 33 (3):147–152. doi:10.1016/S0963-9969(00)00028-4

Kinderlerer JL (1993) Fungal strategies for detoxification of medium chain fatty acids. Int Biodeterior Biodegrad 32(1–3):213–224. doi:10.1016/0964-8305(93)90053-5

Kinderlerer JL, Kellard B (1984) Ketonic rancidity in coconut due to xerophilic fungi. Phytochem 23(12):2847–2849. doi:10.1016/0031-9422(84)83027-7

Kochhar SP (1996) Oxidative pathways to the formation of off-flavours. In: Saxby MJ (ed) Food taints and off-flavours. Springer US, New York, pp 168–225. doi:10.1007/978-1-4615-2151-8_6

Kupiec B, Revell B (1998) Speciality and artisanal cheeses today: the product and the consumer. Brit Food J 100(5):236–243. doi:10.1108/00070709810221454

Lepper AN (2013) Influence of cut, cooking method, and post-mortem aging on beef palatability. Dissertation, North Dakota State University

Lopez-Fandiño R, Carrascosa AV, Olano A (1996) The effects of high pressure on whey protein denaturation and cheese-making properties of raw milk. J Dairy Sci 79(6):929–936. doi:10.3168/jds.S0022-0302(96)76443-3

Macrae AR (1983) Lipase-catalyzed interesterification of oils and fats. J Am Oil Chem Soc 60(2):291–294. doi:10.1007/BF02543502

Marshall VM, Tamime AY (1997) Starter cultures employed in the manufacture of biofermented milks. Int J Dairy Technol 50(1):35–41. doi:10.1111/j.1471-0307.1997.tb01733.x

Martins SIFS, Jongen WMF, van Boekel MAJS (2000) A review of Maillard reaction in food and implications to kinetic modeling. Trends Food Sci Technol 11(9–10):364–373. doi:10.1016/S0924-2244(01)00022-X

McCain GR (1991) Gases. In: Smith J (ed) (1991) Food additive user's handbook. Blackie and Son Ltd, London

Menon AS, Hii CL, Law CL, Suzannah S, Djaeni M (2015) Effects of water blanching on polyphenol reaction kinetics and quality of cocoa beans. In: Proceedings of the International Conference of Chemical and Material Engineering (ICCME) 2015: Green Technology for Sustainable Chemical Products and Processes, Semarang, vol 1699, 29–30 Sept 2015, p 030006. doi:10.1063/1.4938291

Messens W, Van Camp J, Huyghebaert A (1997) The use of high pressure to modify the functionality of food proteins. Trends Food Sci Technol 8(4):107–112. doi:10.1016/S0924-2244(97)01015-7

Micali M, Parisi S, Lagana' P, Minutoli E, Delia S (2009) Il confezionamento in atmosfera modificata. Concetti fondamentali, metodi di disamina dei rischi microbici, applicazioni pratiche. Ind Aliment 48(489):35–43

Mizutani T, Hashimoto H (2004) Effect of grinding temperature on hydroperoxide and off-flavor contents during soymilk manufacturing process. J Food Sci 69(3):SNQ112–SNQ116. doi:10.1111/j.1365-2621.2004.tb13379.x

Morris A, Barnett A, Burrows O (2004) Effect of processing on nutrient content of foods. Cajanus 37(3):160–164

Narayan KA (1997) Biochemical aspects: nutritional bioavailability. In: Taub IA, Singh RP (eds) Food Storage Stability. CRC Press, Boca Raton

Nishita KD, Bean MM (1982) Grinding methods: their impact on rice flour properties (Starch quality). Cereal Chem 59(1):46–49

Ottaviani F (2002) Il metodo HACCP (Hazard analysis and critical control points). In: Andreis G, Ottaviani F (eds) Manuale di sicurezza degli alimenti. Principi di ecologia microbica e di legislazione applicati alla produzione alimentare. Oxoid S.p.A., G. Milanese

Parisi S (2002) La stesura del Piano di Autocontrollo nel comparto agro-alimentare. Chim Ital 13(2):25: 28

Parisi S (2003) Evoluzione chimico-fisica e microbiologica nella conservazione di prodotti lattiero-caseari. Ind Aliment 42(423):249–259

Parisi S (2004) Alterazioni in imballaggi metallici termicamente processati. Gulotta Press, Palermo, Italy

Parisi S (2006) Profili chimici delle caseine presamiche alimentari. Ind Aliment 45(457):377–383

Parisi S (2009) Intelligent packaging for the food industry. In: Carter EJ (ed) Polymer electronics—a flexible technology. Smithers Rapra Technology Ltd, Shawbury

Parisi S (2012) Food packaging and food alterations. Smithers Rapra Technology Ltd, Shawbury

Parisi S (2013) Food Industry and packaging materials: performance-oriented guidelines for users. Smithers Rapra Technology Ltd, Shawbury

Parisi S, Laganà P, Delia S (2006) Il calcolo indiretto del tenore proteico nei formaggi: il metodo
    CYPEP. Ind Aliment 46(462):997–1010
Pilone GJ, Clayton MG, van Duivenboden RJ (1991) Characterization of wine lactic acid bacteria:
    single broth culture for tests of heterofermentation, mannitol from fructose, and ammonia from
    arginine. Am J Enol Vitic 42(2):153–157
Patras A, Brunton NP, O'Donnell C, Tiwari BK (2010) Effect of thermal processing on
    anthocyanin stability in foods; mechanisms and kinetics of degradation. Trends Food Sci
    Technol 21(1):3–11. doi:10.1016/j.tifs.2009.07.004
Priest FG (1999) Gram-positive brewery bacteria. In: Priest FG, Campbell I (eds) Brewing
    microbiology. Springer US. doi:10.1007/978-1-4419-9250-5_5
Rahman MS, Perera CO (1999) Drying and food preservation. In: Rahman MS (ed) (2007)
    Handbook of food preservation, 2nd edn. CRC Press LLC, Boca Raton
Rodriguez-Amaya DB (1997) Carotenoids and food preparation: the retention of provitamin A
    carotenoids in prepared, processed and stored foods. John Snow Incorporated/ Opportunities
    for Micronutrient Interventions (OMNI) Project, Arlington, VA. http://pdf.usaid.gov/pdf_docs/
    Pnacb907.pdf. Accessed 28 Apr 2016
Sams AR (2001) Poultry meat processing. CRC Press, Boca Raton
Sarmadi BH, Ismail A (2010) Antioxidative peptides from food proteins: a review. Peptides 31
    (10):1949–1956. doi:10.1016/j.peptides.2010.06.020
Sherwin ER (1978) Oxidation and antioxidants in fat and oil processing. J Am Oil Chem Soc 55
    (11):809–814. doi:10.1007/BF02682653
Šimko P (2002) Determination of polycyclic aromatic hydrocarbons in smoked meat products and
    smoke flavouring food additives. J Chromatogr B 770(1):3–18. doi:10.1016/S0378-4347(01)
    00438-8
Stadtman ER, Levine RL (2003) Free radical-mediated oxidation of free amino acids and amino
    acid residues in proteins. Amino Acids 25(3–4):207–218. doi:10.1007/s00726-003-0011-2
Sugimura T (1985) Carcinogenicity of mutagenic heterocyclic amines formed during the cooking
    process. Mutat Res 150(1):33–41. doi:10.1016/0027-5107(85)90098-3
Tan CP, Man YC (2002) Recent developments in differential scanning calorimetry for assessing
    oxidative deterioration of vegetable oils. Trends Food Sci Technol 13(9–10):312–318. doi:10.
    1016/S0924-2244(02)00165-6
Tregear A (2003) From Stilton to Vimto: using food history to re-think typical products in rural
    development. Sociol Rural 43(2):91–107. doi:10.1111/1467-9523.00233
Walstra P (2003) Physical chemistry of foods. CRC Press, Boca Raton
Zardetto S, Dalla Rosa M (2006) Study of the effect of lamination process on pasta by physical
    chemical determination and near infrared spectroscopy analysis. J Food Eng 74(3):402–409.
    doi:10.1016/j.jfoodeng.2005.03.029